W9-ADN-721

Combinatorics

Combinatorics

An Introduction

Theodore G. Faticoni
Department of Mathematics
Fordham University
Bronx, NY

A JOHN WILEY & SONS, INC., PUBLICATION

Cover Design: John Wiley & Sons, Inc.
Cover Illustration: courtesy of Theodore G. Faticoni

Published by John Wiley & Sons, Inc., Hoboken, New Jersey.
Published simultaneously in Canada.

Library of Congress Cataloging-in-Publication Data:

Faticoni, Theodore G. (Theodore Gerard), 1954–
Combinatorics : an introduction / Theodore G. Faticoni, Department of Mathematics, Fordham University, Bronx, NY.
pages cm
Includes bibliographical references and index.
ISBN 978-1-118-40436-2 (hardback)
1. Combinatorial analysis. I. Title.
QA164.F38 2012
511'.6—dc23 2012025751

Printed in the United States of America.

10 9 8 7 6 5 4 3 2 1

To our mother, Margaret Faticoni.

My sisters and I are one for our Mom's efforts.

Contents

Preface

As I read the current textbooks on finite or linear mathematics, I am struck by the superficial way that counting problems or combinatorics are handled. Counting is treated as a methodical or mechanical thing. The student is asked to memorize a few important but unenlightening algorithms that will always tell us the number of ways that someone can choose and arrange her outfits for the week.

Furthermore, the examples that are given use so much of reality that the student has more to learn about electronic components, failed tests, and card games than they do about counting in mathematics. Whatever happened to problems that emphasized their mathematical content and left a knowledge of science and gaming to other departments? I understand that, to some, mathematics is best when it is used in applications. But why are we giving up on teaching mathematical content in favor of these other subjects?

Moreover, the why of it all, the justification, the beauty of proof has left these courses entirely. There is no explanation as to how the fundamental formulas are derived, and there is no rationalization as to how certain formulas are formed. The exercises that are given in modern texts are just slight variations on the examples worked out in the chapter. And in my opinion, the chapter examples are mostly uninspiring.

This book is aimed at college students, teaching assistants, adjunct instructors, or anyone who wants to learn a little more elementary combinatorics than the usual text contains. This book might also be used as a supplement to the existing text for a finite mathematics course or to supplement a discrete mathematics course, which several curriculums require.

The purpose of this book is to give a treatment of counting combinatorics that allows for some discussion beyond what is seen in today's texts. We will discuss and justify our formulas at every turn. Our examples will include, after the most elementary of applications, some ideas that do not occur in other texts on the market at this time. The applications never get beyond the use of Venn diagrams, the inclusion/exclusion formula, the multiplication principal, permutations, and combinations. But their uses are clever and at times inspiring.

For example, we do some poker hand problems that are not seen in modern texts, we count the number of bracelets that can be made with $n > 1$ different colored beads, and we count the number of derangements of $\{1, \ldots, n\}$. We do this without any more than the elementary tools for counting. We then consider some probability problems by doing some elementary counting. But we show some very surprising, mathematically precise consequences of a trained approach to the subject.

A second theme within this book is that the case-by-case method for solving problems is emphasized. Of course we use a formula when needed, but when it comes time to derive a formula, we have decided to consistently give the case-by-case approach to the problem. In this way we are asking the student/reader to think mathematically and in exactly the same way from problem to problem throughout the book. Perhaps this is what the students will take with them when they leave the course. They will misremember the applications for the permutation formula, but they might remember how to break a problem into pieces in order to solve it.

The book is a series of short chapters that cover no more than one topic each. We cover such topics as logic and paradoxes, sets and set notation, power sets and their cardinality, Venn diagrams, the multiplication principal, permutations, combinations, problems combining the multiplication principal, problems combining permutations and combinations, problems involving the complement rule, *at least*, and *at most*. We cover derangements, elementary probability, conditional probability, independent probability, and Bayes' Theorem. We close with a discussion of two dimensional geometric simplex algorithm problems, showing that the traditional geometric method breaks down in the case where the variables take on only integer values. In other words, the method breaks down in every example done in the modern finite mathematics texts.

There are plenty of worked examples, as I want to do the work for the reader, and there is a short list of homework exercises. The examples given can also be used by an instructor or a teaching assistant to gain a higher level of understanding of the subject than the current texts offer, thus providing the instructor with an overview of the subject that the student does not possess. This can aid the classroom situation since, as I believe, we do a better job of teaching when we teach from a higher point of view in the delivered subject. The instructor then has a professional confidence that (s)he can solve any problem that comes up in class.

The fact that the book is salted with explanations as to why certain formulas exist helps the student and the instructor understand what they are doing. This is different from the rote memorization that many texts on this subject require. In this book the justification for the formulas is also there.

With this approach to the subject and to my readers, I believe I have found a gradual, understandable path that will bring a college student to a discussion of a subject on combinatorics and probability that is more advanced than any of the topics covered in the current texts on finite mathematics.

Theodore G. Faticoni
Department of Mathematics
Fordham University
Bronx, New York 10458
faticoni@fordham.edu

Chapter 1

Logic

There are several kinds of logic in mathematics. The one based in the construction of Truth tables is called *formal logic.* This is the logic used in computer science to design and construct the guts of your computer. And then there is Aristotle's logic. This is the logic used to make arguments in court or when arguing informally with another person. This is the logic used to prove that something is, or to prove that something is not. This is the logic used to examine combinations of any of the mathematical ideas encountered in this text. While we will examine formal logic and the logic of sets and functions, we will be most interested in Aristotle's *logic of the argument* in this chapter and throughout the rest of the text.

Oh, and there will be no need for a calculator in this book. I have made an effort to emphasize the important mathematical content in this book, not the superfluous, tedious practice of arithmetic. Arithmetic is important when you work with money, but in more challenging mathematical problems it only gets in the way. So cradle your electronic toy if you need to, but there will be almost no use for it as we do our counting.

1.1 Formal Logic

Formal logic is just a series of tables describing how the words *and, or, not* are defined. There is nothing illuminating with this approach, but it does match the operations of the inner workings of your computer. We will minimally justify the tables used here. We will just write them down and show how they agree with your use of the words in your language.

These tables define logic. Not just in English, the language that this book is being written in, but they describe logic in *every* language on earth. If you are reading a Mandarin Chinese translation of this book, then the logic presented here will still be the logic of your language. It is also the binary language in which the software in your computer is written. Take time to savor that thought. Logic as it is applied to languages and computers is universal. Logic is thus common to all forms of communication, analogue or digital.

To begin with we need to know what the logical operations are and what they operate on. *Logic operates on statements*, and ordinarily we will use the letters P, Q, and R to denote the statements that we we are working on. These statements can take on the *logical states* T (for True) and F (for False).

You already have an intuitive understanding of what it means for a statement to be True or False. You know that *The sky is blue* is True on earth, and you know that *You and I are human* is a True statement. *You have five dollars* might be True right now, but it might be False come late Friday evening. Of course *It is raining* is a False statement on a sunny day over my home, but it might be a True statement for you where you live. So let us assume that we know what T and F mean in this context.

The first logical operation that we will investigate is the operation *not*. The *not* operation takes a statement P and changes or negates its logical states. It changes T to F and F to T. Its *Truth table*, the table that lists the logical states of the *not* operation, follows.

P	not P
T	F
F	T

This is just a tabular way of defining what *not* is. Notice that according to the table, if P is T then not P is F, and if P is F then not P is T. As we said, *not* changes a statement's logical state to the complementary logical state.

EXAMPLE 1.1.1 1. If P is the statement *The sky is blue on earth*, then not P is the statement *The sky is not blue on earth.* We have negated P and changed its logical state from T to F.

2. If P is $1+2=3$ then not P is the statement $1+2\neq 3$. Again the logical state of P has been changed by an application of *not* from T to F.

Because of the nature of the word *not*, two consecutive applications of the operation *not* to P will leave the logical states of P unchanged. For lingual reasons we let not not P = not(not P). In tabular form the compound operation *not not* is written as follows.

P	not P	not(not P)
T	F	T
F	T	F

Notice that if P is T then not P is F, and then not(not P) is T, giving not(not P) the logical states of P. You know this as a *double negative* from your English class.

EXAMPLE 1.1.2 1. If P is *The sky is blue on earth*, then the double negative not(not P) is the awkward sentence *It is False that the sky is not blue on earth.* Your language skills compel you to avoid the double negative and just write *The sky is blue on earth.*

2. Suppose P is *I think this is wrong.* Then not P is *I think this is not wrong*, and not(not P) is the very awkward *I don't think that this is not wrong.* You would be advised by your language teacher to avoid the double negative and just say *I think this is wrong.* The statements P and not(not P) are written with different words, but logically they express the same meaning.

Thus, by applying the logic of the operator *not* to a lingual double negative, we can avoid the double *not.*

Throughout this discussion, suppose that we are given statements P, Q. Several logical operations allow us to compare the logical states of P, Q by combining them.

For instance, we can combine statements P, Q using the *and* operation. This is the *and* that you use all of the time when you write. When applied to P, Q the *and* operation yields the statement "P and Q". This is just the compound statement formed by combing P, Q with the conjunction *and* from English.

EXAMPLE 1.1.3 1. If P is *The sky is blue on Earth* and if Q is *You are a man* then "P and Q" is the statement *The sky is blue on Earth and you are a man.*

2. If P is *This is wrong* and if Q is *These are red* then "P and Q" is *This is wrong and these are red.*

The logical states of P and Q are closely related to the way that the word *and* behaves in language. Thus the logical state of P and Q is T (True) exactly when both P and Q are T. In every other instance, "P and Q" is F (False). Put another way, if one or more of the logical states of P, Q are F (False) then the statement "P and Q" is a Falsehood, its logical value is F.

In the form of a Truth table the *and* operation is diagrammed as follows:

P	Q	P and Q
T	T	T
T	F	F
F	T	F
F	F	F

The first row states that if both P, Q have logical state T then the conjunction "P and Q" also has logical state T. Once we know that the right hand entry of the first line in the table is T then the rest of the rows follow as F.

EXAMPLE 1.1.4 1. If P is *I am a human being* and if Q is *I am sitting in my chair* then "P and Q" is T exactly when *I am a human being* is T and *I am sitting in my chair* is T. Any other combination of T's and F's for P, Q will produce a logical state F for "P and Q".

2. If P is *The sky is red over me* and if Q is *The ground is dry beneath me* then the logical value of "P and Q" is F if we are

on Earthsince the sky is not red there. If we are on Mars then the logical value of "P and Q" is T because the sky is red and the ground is dry on Mars.

Another way to combine statements is through the use of the conjunction *or*. The use of *or* in logic is denoted by the operation *or*. Thus, statements P, Q are combined to form the conjunctive statement "P or Q", which is read just like the *or* statements that you read and write.

The compound statement "P or Q" has logical state T exactly when one or more of the statements has logical state T. But it might be easier to remember how *or* behaves with False statements. When the logical states of both P and Q are F then "P or Q" has logical state F, and this is the only case in which the logical state of "P or Q" is F.

We will always use the *inclusive or* here so that the statement "P or Q" includes the case where both P, Q have logical state T. That is, we we read "P or Q" as *P, Q, or both P and Q.*

EXAMPLE 1.1.5 1. If P is *The river is wide* and if Q is *The water is cold* then "P or Q" is read as *The river is wide or the water is cold.* Since "P or Q" is T when either P, Q has logical state T, the compound statement *The river is wide or the water is cold* has logical state T if the river is wide.

2. *The river is wide or the water is cold* is T if we are talking about the Missouri River and its waters are cold. *The river is wide or the water is cold* is T if we are talking about the Missouri River and the water we are talking about is in my coffee.

3. Let P be the statement *All is nothing* and let Q be the arithmetical statement $1 + 1 = 3$. Both P and Q have logical state F, so that "P or Q" has logical state F. Since both P, Q have logical state F then "P or Q" has logical state F.

The next logical operations, called *DeMorgan's laws*, show us how the logical operations *and, or, not* combine with each other. Simply put, *DeMorgan's laws* are lingual ways of simplifying a sentence that uses *and* , *or*, and *not* is a more complex manner.

Given statements P, Q then *DeMorgan's laws* are written as

$$\begin{aligned} \text{not}(P \text{ or } Q) &= (\text{not } P) \text{ and } (\text{not } Q) \\ \text{not}(P \text{ and } Q) &= (\text{not } P) \text{ or } (\text{not } Q). \end{aligned}$$

Notice that in our use of DeMorgan's Law, the distribution of the *not* operator changes *or* to *and*, or it changes *and* to *or*. Compare this to the following lingual examples of uses of DeMorgan's laws. When read properly, you will see that the symbolism we use here is the same as our use of *and, or, not* above.

We will use parentheses to emphasize a statement's meaning, so that there is no confusion as to what word modifies what phrase.

EXAMPLE 1.1.6 1. The statement

(The river is not wide) or (the water is not cold)

is equivalent to the statement

It is not True that (The river is wide and the water is cold).

Complex to be sure, but that is the purpose behind DeMorgan's laws. It will take a complicated statement and make it easier to read.

2. The statement

(This is not a king) and (this is not a queen),

is equivalent to the statement

This is not (a king or a queen).

3. The statement

This box does not contain (a red and a yellow crayon),

is equivalent to

(This box does not contain a red crayon) or
(it does not contain a yellow crayon).

EXAMPLE 1.1.7 1. Let P be the statement that *This is a king* and let Q be the statement that *This is a queen.* The statement "not(P or Q)" is also written as

It is False that (this is a king or a queen),

while "(not P) and (not Q)" is written as

(This is not a king) and (this is not a queen).

Which do you prefer? Logically they both mean the same thing.

2. Let P be the statement that *This box contains a red crayon* and let Q be *This box contains a yellow crayon.* Then "not(P and Q)" is written as

It is False that (this box contains a red and yellow crayon),

while its equivalent formulation "(not P) or (not Q)" under De-Morgan's laws is

(This box does not contain a red crayon) or
(this box does not contain a yellow crayon).

1.2 Basic Logical Strategies

We will make exclusive use of logical arguments due to Aristotle some 500 years B.C. They are the basis for every intelligent conversation and every legal argument made since.

The first logical observation is that one statement always has a logical state of F.

The statement "P and (not P)" is a universal Falsehood.

No matter what the logical state of P is, "P and (not P)" is a Falsehood.

To see this, notice that because *not* changes logical states, at any time either P or not P is F. Thus the *and* statement "P and (not P)"

has logical state F. The Truth table for "P and (not P)" is then given as follows:

P	not P	P and (not P)
T	F	F
F	T	F

Observe that the right-hand column of the table is made up of F's. Thus, the statement "P and (not P)" is a Falsehood.

EXAMPLE 1.2.1 1. Let P be *The sky is blue.* Then *(the sky is blue) and (the sky is not blue)* is a Falsehood.

2. Let P be *This statement is True.* Then "P and (not P)" is the statement *This statement is True and this statement is not True*, and this is a Falsehood.

3. Let P be *There is a mountain.* Then "P and (not P)" is *(There is a mountain) and (there is no mountain)*, which is a Falsehood. So is *First there is a mountain, then there is no mountain, then there is.*

We continue our discussion of logical arguments. Given statements P, Q, the statement "P implies Q" is called an *implication*, and it is symbolically written as

$$P \Rightarrow Q.$$

The statement P is called the *premise of the implication* and Q is called its *conclusion.*

The logical states of $P \Rightarrow Q$ are determined by one line of explanation.

> If your argument is correct then Truth leads to Truth.

In other words, if your argument is T and if your premise P is T then your conclusion Q is T. Every other logical state of $P \Rightarrow Q$ follows from this boxed statement.

Note that line one of the following Truth table for "$P \Rightarrow Q$" is logically equivalent to the boxed statement above.

P	Q	$P \Rightarrow Q$
T	T	T
T	F	F
F	T	T
F	F	T

Let us fill in the remaining Truth values for this table. Let P and Q be statements and consider "$P \Rightarrow Q$". We will show how a few simple Truths about argument discovered by Aristotle can be used to fill in the Truth table for the *implication.*

EXAMPLE 1.2.2 We will continually refer to the Truth table for "$P \Rightarrow Q$".

1. Because Truth implies Truth when the argument is correct,

> If your argument is correct (T), and if P is T then Q is T.

This is why line 1 is

P	Q	$P \Rightarrow Q$
T	T	T

.

2. Since Truth implies Truth when the argument is correct,

> Your argument is False if P is T and Q is F.

This is why line 2 of the Truth table is

P	Q	$P \Rightarrow Q$
T	F	F

.

3. Since any argument begun with a False premise is correct, we can write

> Your argument is T if P is F.

This is why lines 3 and 4 of the Truth table are

P	Q	$P \Rightarrow Q$
F	T	T
F	F	T

.

The column under Q is the list of all possible logical states for Q in the Truth table for "$P \Rightarrow Q$".

4. Since a False premise leads to either a True or False conclusion,

> Your conclusion is ambiguous if P is F.

This is why lines 3 and 4 of the Truth table are

P	Q	$P \Rightarrow Q$
F	T	T
F	F	T

.

The column under Q completely describes an ambiguous conclusion Q. The T's under "$P \Rightarrow Q$" result from the part 3.

Let us put this implication to work in some elementary arguments.

EXAMPLE 1.2.3 1. Here is a Greek classic. We will use Example 1.2.2(1). Begin with P : *Socrates is a man.* The conclusion will be Q : *Socrates is mortal.* The implication $P \Rightarrow Q$ is *If Socrates is a man then Socrates is mortal.* Since the implication $P \Rightarrow Q$ is correct, and since the Truth of the premise P implies the Truth of the conclusion Q, Socrates is mortal.

2. The premise is P : *I stand on dry land on earth*, and the conclusion is Q : *The sky above me is blue.* The implication is *If I stand on dry land on Earth then the sky above me is blue* is True. Since P is True, and since Truth leads to Truth, Q is True.

3. The premise is P : *Digital technology is like pockets*, and the conclusion is Q : *We have had digital technology for hundreds of years.* The implication is "$P \Rightarrow Q$" *We have had pockets for hundreds of years.* Let us assume that the premise P is True. Since Q is Falsehood, the implication "$P \Rightarrow Q$" has logical state F. But

if we assume that the premise P is False, then Q is still False, but the implication "$P \Rightarrow Q$" is True.

4. Under what conditions will P in part 3 lead us to a True conclusion Q? Have fun with this one.

1.3 The Direct Argument

This formal manipulation of statements is not exactly what we are interested in for this chapter. It is good to know that an argument has logical state T or F, but it is better to know how we can use the implication to correctly deduce a conclusion.

The first line T, T, T of the Truth table for $P \Rightarrow Q$ can be restated as *If our argument is correct then Truth leads to Truth,* or in other words, *If the premise is True and if the argument is correct then the conclusion is True.* This form of argument is called the *direct argument.* It is not new to you since you unconsciously use direct arguments in your everyday life.

EXAMPLE 1.3.1 1. The premise is P : *The sky is not blue* and the conclusion is Q : *We are not on earth.* A correct argument is

If the sky is not blue then we are not on earth.

Conclude that the conclusion Q is True.

2. Something more mathematical begins like this. The premise is $P: 1+1=2$. Argue correctly as follows:

$1+1=2$
If we add 1 to both sides of $1+1=2$ then $1+1+1=2+1$.
If $2+1=3$ then $1+1+1=3$.
The conclusion $Q: 1+1+1=3$ is then True.

A chain-like form of argument shows us the structure inherent in longer arguments called *transitive property.* These longer arguments are what people make when they logically move from one idea to the next. Basically, the *transitive property of implications* is a way to leap from two or more implications to one implication. Hence

If $P \Rightarrow Q$ and if $Q \Rightarrow R$ then $P \Rightarrow R$.

A series of implications and the transitive property provide us with a method for arguing efficiently with many implications. This series of implications is called the *transitive argument.*

> Assume the Truth of the premise P.
> Show that $P \Rightarrow Q$ is True
> Show that $Q \Rightarrow R$ is True
> Conclude the Truth of R.

To justify that this column forms an argument that we can use to deduce R from P, we will argue lingually.

Proof: Assume the Truth of P. If $P \Rightarrow Q$ is True then by the Direct Argument Q is True. If $Q \Rightarrow R$ is True then by the Direct Argument we conclude the Truth of R. Therefore, our transitive argument concludes the Truth of R from the Truth of P.

Let us review what we just argued in terms of True statements. We begin with a True statement P. The assumption is that $P \Rightarrow Q$ and $Q \Rightarrow R$ are True, which allows us to make a correct transitive argument

$$P \Rightarrow Q \text{ and } Q \Rightarrow R \text{ implies } P \Rightarrow R.$$

From the Truth of P and the Truth of $P \Rightarrow R$ we use the Direct Argument to conclude the Truth of R.

In a later section we will argue as we did above and in greater detail, thus producing three more argument forms.

EXAMPLE 1.3.2 This example shows how the above discussion can be applied to longer arguments.

a) The premise is P: $10 < 2^{10}$.
b) $P \Rightarrow Q$: Because $10 < 2^{10} = 1024$ then $11 < 2^{10}$.
c) $Q \Rightarrow R$: Because $11 < 2^{10}$ then $11 < 2 \cdot 2^{10} = 2^{11}$.
d) Conclude R: $11 < 2^{11}$.

Using this iterated form of argument people form longer and more complicated arguments, which allows them to perform more complicated intellectual tasks. These tasks could be just a way of adding numbers, or it could be the design of your computer's software, or it could be that the arguments take the arguer to intellectual places that no one had conceived before. The lesson to learn here is that, while the tabular thinking of logic is good for some tasks, there will always come a time in problem solving when we must use argument and a more enlightened form of thinking if we are to make progress on hard problems.

REMARK 1.3.3 When your computer operates it is working its way through a very long and tedious argument based on the very simple *binary logic* introduced in this section. The steps in the computer's argument are mechanical, a form of arithmetic completed by a machine. The men and women who designed this computer had to think through the *binary logic* during the implementation phase of the software.

However, for the men and women who put the larger internal logical parts of the computer together in the design phase, the problems encountered could not be solved with a simple manipulation of *binary logic*. They had to think creatively through the problems presented to them by the design phase. These solutions would often include a leap of the imagination that could not be anticipated when the design for the computer was initially proposed. The logical problems yet to come will require those leaps of the imagination before we can solve our problems.

1.4 More Argument Forms

Converse Statements

The implication $P \Rightarrow Q$ comes with what is called its *converse*.

The converse of $P \Rightarrow Q$ is $Q \Rightarrow P$.

Let us write down the Truth table for $Q \Rightarrow P$ and compare it to $P \Rightarrow Q$.

P	Q	$P \Rightarrow Q$
T	T	T
T	F	F
F	T	T
F	F	T

P	Q	$Q \Rightarrow P$
T	T	T
T	F	T
F	T	F
F	F	T

As you can see, the implication and its converse do not have the same Truth table. The logical state of the implication $P \Rightarrow Q$ in the third row is T, while the logical state of the implication $Q \Rightarrow P$ in the third row is F. Thus the converse implication $Q \Rightarrow P$ can have logical state F even when $P \Rightarrow Q$ has logical state T. For this reason, the converse cannot be used as a True statement even when the original implication is True. Hence *all* are forewarned to avoid the classic error of using the converse of an implication to advance an argument.

EXAMPLE 1.4.1 These examples show that we cannot interchange the implication with its converse. They will have different logical states.

1. Let P be the T statement *The sky is blue*, and let Q be *The world is flat.* Then "$P \Rightarrow Q$" is F.

The converse of "$P \Rightarrow Q$" is the statement "$Q \Rightarrow P$:" *If the world is flat then the sky is blue.* Since its premise Q is F, "$Q \Rightarrow P$" is T. Thus the implication is False while the converse is True, and we cannot exchange them in arguments or conversation.

2. The implication is *If today is Monday then my schedule is clear* and its converse is *If my schedule is clear then today is Monday.* The implication may be True, but the converse is False since my schedule is clear on Sunday.

Contrapositive Statements

Suppose that we consider the implication $P \Rightarrow Q$, assuming that it is T. If Q is F then the Truth table for $P \Rightarrow Q$ shows us that P is also F. Thus, a False premise Q implies a False conclusion P. This is an important implication known as the *contrapositive.*

not $Q \Rightarrow$ not P.

When one writes out the Truth table for the implication and its contrapositive, a curious thing occurs. This Truth table reveals that the two arguments have identical Truth tables.

P	Q	$P \Rightarrow Q$	not $Q \Rightarrow$ not P
T	T	T	T
T	F	F	F
F	T	T	T
F	F	T	T

Notice that the rightmost two columns are identical lists of T's and F's. This is completely different from what we found with the converse. The table shows that

The implication and its converse are logically equivalent. One can be substituted for the other without loss of Truth.

In other words, the statements "$P \Rightarrow Q$" and "not $Q \Rightarrow$ not P" are both True for the same logical values of P and Q.

EXAMPLE 1.4.2 1. The implication *If the sky is not blue then this is not earth* has as contrapositive *If this is earth then the sky is blue.* The implication and its contrapositive are making the same logical statement about the sky.

2. The implication *If my GPS is working then I am not lost* has contrapositive *If I am lost then my GPS is not working.* Notice that both the implication and its contrapositive are making the same logical statement, assuming I always use my GPS.

3. The implication *If my spell-check program is running then I do not misspell all the time* has contrapositive *If I misspell all the time then my spell-check program is not running.* Notice that both the implication and its contrapositive make the same logical statement about a man who cannot spell without technological help.

Counterexamples

The next form of argument does not use T's and F's. It is strictly lingual.

Let P be a statement. A *counterexample to* P is an example that is in logical conflict with the content of P. The existence of a counterexample to P proves that P is False.

The idea behind the proof by counterexample is this. If I claim that P : *All colors are white* is True then you can disprove my claim by *producing some color that is not white.* One non-white color will do. I choose red. With the existence of the color red you have refuted my claim. You have proved that *All colors are white* is a Falsehood.

In the very same manner, we can disprove any statement that asserts that all of the X's in the world are short Y's. All we need do is find a counterexample X that is not a short Y.

The *proof by counterexample* can be summed up as follows:

> The statement *All X's have property Y* is **disproved** by a counterexample of an X that does not have property Y.

These proofs by counterexample all proceed in the same way. Producing just one X that *does not have* quality Y is enough to kill the claim that *All X's have property Y*.

EXAMPLE 1.4.3 1. The claim is P : *All integers are even.* To refute the claim you produce counterexample 3, which is not even. This counterexample refutes the claim that *all integers are even.* You have thus disproved the claim that *All integers are even.* Hence *Some integer is odd.*

2. The claim P : *All people are Truth sayers* claims that *every person tells the Truth at all times.* Your counterexample to refute the claim is the known Falsehood *"1=0"*. Having uttered a False statement, the claim P is disproved.

3. The claim P : *All people are liars* claims that *every person will lie at all times.* Your counterexample to refute the claim is

the known Truth "$1 \neq 0$". Having uttered a Truth, the claim P is disproved.

4. Your claim is P : *All statements are False.* I state that *The number line has no end.* My stated Truth is a counterexample that refutes or disproves the claim.

5. You claim P : *There are no interesting positive integers.* I argue thusly: in that case, there is a least or minimum non-interesting integer, call it x. I find it interesting that there exists such a number, and so I find x interesting. This interest in x is a counterexample to the claim.

6. Someone claims that *Nothing in this world is interesting at all.* I argue that the lack of interesting facts in this world is interesting to me. This shows that something is interesting to someone, which is a counterexample to the claim.

1.5 Proof by Contradiction

In this section we will show that a certain kind of statement is always a Falsehood. These statements are common among amateur mathematicians who do not fully understand the logical ideas that we have been examining in this chapter. We will show that these Falsehoods have similar proofs even though they do not look alike. These proofs are so similar that one proof will be used on one statement by simply replacing certain words in a previous proof. The examples below will make this clear.

We begin with a logical problem that comes from ancient Greece circa 600 B.C.

EPIMENIDES PARADOX 1.5.1 The Cretan Epimenides steps onto a stage in Athenian College and proudly speaks his three lines.

> All Cretans are liars.
> All statements made by Cretans are False.
> I am lying.

The Greek scholars proceed to determine the logical value of *I am lying* in a manner that we will read promptly. They decide that *I am lying* is neither a lie nor the Truth, an intolerable logical situation in any age. We ask for an explanation as to how *I am lying* could lack a logical state.

It is traditional to state that *All Cretans are liars* rather than *All statements made by Cretan's are False.* We will consistently use the longer version for now. To begin our modern approach to the Epimenides Paradox, we will show that the premise for the scholar's discussion is a Falsehood.

THEOREM 1.5.2 All statements made by Cretan's are False *is a Falsehood.*

Proof: This is an obvious proof of the theorem. Because the lying Cretan Epimenides speaks *All statements made by Cretans are False*, the statement is itself a lie. This completes the proof.

One might also disprove *All statements made by Cretans are False* with a counterexample. The first one that comes mind is the Truth *Crete is an island.* Yet another proof that *All statements made by Cretans are False* is False will be used as a template for subsequent proofs in this section.

REMARK 1.5.3 Let R be a statement. Any argument that begins by *assuming something* for the sake of contradiction, and that then concludes both R and its logical negation not R, has concluded a Falsehood called a *contradiction.* Because Truth leads to Truth when the argument is correct, we have proved that the something we assumed initially is a Falsehood. We will make extensive use of this form of proof called *proof by contradiction.*

A *self-referential statement* is a sentence that refers to itself in its lingual content. Statements like *This statement is True*, or *This statement is too long*, or, my favorite, *This statement is self-referential.* A statement labeled with a Q in this section is called a *Q-statement.* Q-statements are examples of *self-referential statements.*

THEOREM 1.5.4 All statements made by Cretans are False *is a Falsehood.*

Proof: Assume for the sake of contradiction that *All statements made by Cretans are False*, and consider the statement

> Q: This statement when spoken by a Cretan is not False.

The content of Q states that the statement Q is not False, so we have deduced the statement *Q is not False.* By hypothesis Q, because it is spoken by Epimenides, is False, so *Q is False* is deduced. But R: *Q is False* and its logical negation not R: *Q is not False* form a contradiction. Hence, our premise *All statements made by Cretans are False* is itself a Falsehood, which completes the proof.

Returning to the ancient paradox, we proceed from the False premise *All statements made by Cretans are False.* Thus we can deduce many things, but we have no means of deciding the Truth of those deductions.

Let us quickly review how we argued above. We assumed that the statement P : *All statements made by Cretan's are False* is True. We then deduced the two statements R: *Q is False* and its logical negation not R: *Q is not False*, a contradiction. Since the conclusion is False, we deduce that our premise *All statements made by Cretans are False* is a Falsehood.

EXAMPLE 1.5.5 Let us give a logical analysis of *I am lying* in the context of the Epimenides Paradox above.

We claim that we have deduced that *I am lying* is True, but the Truth is that we cannot identify the logical state of *I am lying.* Beginning with a Falsehood the way we did makes any analysis of the logical state of *I am lying* within the Epimenides Paradox impossible. This illustrates just how badly facts can be distorted when an argument proceeds from a False premise.

That was fun. I hope you derive many hours of pleasure from thinking about the Epimenides Paradox. This logical puzzle demonstrates that if you start with a Falsehood, as we did, then you cannot decide the actual logical state of your conclusion. Now let us consider variations on Epimenides.

EXAMPLE 1.5.6 A Truth sayer is a person who says nothing but the Truth. We will show that P: *All people are Truth sayers* is a Falsehood.

Proof: The method of proof utilized here is the proof by counterexample. One counterexample is produced when I speak the Falsehood $1 = 0$, which is contrary to the statement P. Just for the fun of it, try one of your own counterexamples.

Here is a slight variation on the above example that again shows how we can use a Q *statement.*

EXAMPLE 1.5.7 Assume for the sake of contradiction that *All people are Truth sayers*, and assume that some person speaks the Q-statement

Q: I am lying.

Since we are assuming that *All people are Truth sayers*, Q *is True.* But by its content Q is a lie, or equivalently Q *is not False.* We have thus deduced Q *is True* and its logical negation, a contradiction. Hence *All people are Truth sayers* is a Falsehood.

EXAMPLE 1.5.8 P: *All statements are True* is shown to be a Falsehood by the use of the counterexample and False statement $1 = 0$.

What follows is an alternative proof that P: *All statements are True* is False, by using the indirect argument and a Q *statement.*

THEOREM 1.5.9 All statements are True *is a Falsehood.*

Proof: For the sake of contradiction assume the statement P : *All statements are True*, and consider the statement

Q: This statement is False.

Since P is True it follows that *Q is True.* But the content of Q-states that *Q is False.* We have deduced *Q is True* and its logical negation *Q is False*, a contradiction. This contradiction proves that our assumed statement P: *All statements are True* is a Falsehood.

Now let us examine several universal statements whose logical state cannot be resolved with a simple counterexample.

THEOREM 1.5.10 All opinions are valid *is a Falsehood.*

Proof: For the sake of contradiction assume *All opinions are valid*, and consider the statement

Q: This opinion is not valid.

Because Q is an opinion, our assumption asserts that *Q is valid.* But the content of Q asserts that *Q is not valid.* We have thus deduced the statement *Q is valid* and its logical negation *Q is not valid*, a contradiction. Hence *All opinions are valid* is a Falsehood. This completes the proof.

You might try to prove that *All opinions are valid* by counterexample, but I do not suggest it. Here is the problem if you try this method of attack in an argument about valid opinions.

EXAMPLE 1.5.11 Suppose you are in a debate about the logical state of *All opinions are valid.* Your correct approach would be to produce an opinion that you claim is not valid. Your debate opponents would then claim that you have produced a valid opinion.

The difficulty with this argument by counterexample is that no one knows a *precise* definition of the term *valid.* No one knows because *valid* is usually given several definitions. Some of these are *having enough vowels, having the right number of words, a professional's conclusion about a scientific argument, an irrational response to the use of opinions.* Therefore, no one knows a precise definition of *valid opinion.*

Without those definitions your debate opponents could claim that every counterexample you put forth is actually a perfectly valid opinion. Opinions claimed by your debate opponents to be *valid* might include statements like $1 = 0$, *you do not exist, and there is no universe.* Since you do not know what *valid* means, anyone arguing with you could legitimately claim that your statements are perfectly valid opinions.

The proof in Theorem 1.5.10 avoids the definition of *valid* by deducing the statement *Q is valid* and its logical negation *Q is not valid*, a contradiction. Therefore, whatever the definition of *valid* is, this contradiction proves that our premise *All opinions are valid* is a Falsehood.

The following example considers the logical state of the statement *All is known.* The statement itself has a problem, as no one has ever written a convincing explanation of what *known* means in this context. Does it mean that we know the logical states of everything, or does it mean that we know the meaning of everything? As yet, no one has answered this question in a lingually professional manner. Nor has anyone realized that this use of the word *All* creates a logical and temporal conflict. If *All is known* then when did you know what *All* means. Does it mean that *All of everything is known* or did it mean that the word *All* is known? At present no one has given a cogent explanation as to why any of these questions can be ignored.

Our next example demonstrates that we do not need to know what *All is known* means.

THEOREM 1.5.12 All is known *is a Falsehood.*

Proof: The proof we use here is exactly the proof used in the previous example where we proved that *All opinions are valid* is False. We will simply replace *opinion* with *statement* and *valid* with *known.* This is surprising physical evidence that *All opinions are valid* and *All is known* are actually the same type of Falsehood.

For the sake of contradiction assume *All is known*, and consider the statement

Q: This statement is not known.

We assumed that all is known, so we deduce Q *is known.* Moreover, the content of Q asserts that Q *is not known.* We have thus deduced the statement Q *is known* and its logical negation Q *is not known*, a contradiction. Therefore, *All is known* is a Falsehood.

The above examples illustrate a general form of statement and argument that can be used to prove that an abstract idea is actually a Falsehood. Let L be a list of *qualities of statements* that contains and that is not restricted to the values *True, known, valid, complicated, assumed, hard to understand.* Fix a quality $Y \in L$ of statements.

Let X be a set of statements that include

Q: This statement in X does not have quality Y.

Evidently, the assertion *All statements in* X *have quality* Y and its abbreviated form

All in X are Y

are logically equivalent. Let us examine the logical state of *All in* X *are* Y. Note that the proof of the following theorem depends on the introduction of a Q-statement in a manner identical to the above proofs.

THEOREM 1.5.13 *Let* X *be a set of statements that contains* Q. *Then* All in X are Y *is a Falsehood.*

Proof: By hypothesis, $Q \in X$. For the sake of contradiction assume *All in* X *are* Y. Because $Q \in X$, our assumption implies that *Q has quality Y*. Moreover, the content of *Q* asserts that *Q does not have quality Y*. We have thus deduced the statement *Q has quality Y* and its logical negation *Q does not have quality Y*, a contradiction. Hence *All in X are Y* is a Falsehood, which completes the proof.

A *perfect logician* is a person who knows all of logic. Let us use our methods to deduce that perfect logicians do not exist.

THEOREM 1.5.14 *There are no perfect logicians.*

Proof: For the sake of contradiction assume that there is a perfect logician, and consider the statement

Q: This statement is not known to some perfect logician.

The self-referential statement Q is a statement of logic, so we deduce the statement *Q is know to every perfect logician.* On the other hand, the content of Q states that *Q is not known to some perfect logician.* We have thus deduced *Q is known to every perfect logician* and its logical negation *Q is not known to some perfect logician,* a contradiction. Therefore, there are no perfect logicians, which completes the proof.

Let us apply our work on *perfect logicians* to a logical puzzle that some consider to be the hardest ever fashioned.

EXAMPLE 1.5.15 At the time of this writing, *The World's Hardest Logic Puzzle* has been a popular stop for those who surf the web. The puzzle begins with 200 perfect logicians on an island, 100 of them are blue eyed, and 100 of them are brown eyed. The problem is that these perfect logicians must determine their eye color through the use of logic alone. When they do, they can leave the island, but not before.

That's it. That is all that we assume in this version of the puzzle. There are some Internet versions of this puzzle that include much more detail than this version, but they and their solutions follow from our solution given below. In other words, once we solve this problem then we can solve any other version of it. Indeed, the manner in which we solved the puzzle makes any further logical investigation unnecessary.

The solution is that the problem begins by assuming that there are 200 perfect logicians, while we have proved that there are *no* perfect logicians. Thus the problem proceeds from a False premise. You can therefore deduce anything you want, but you have no way of knowing which deduction is True. Thus you might deduce using 99 theorems that 200 people leave the island Friday, or you might deduce in a few lines that 200 people leave the island instantly, or you might deduce that seven of them never leave the island.

But we cannot know the logical state of any of these deductions, because we proceed from the False premise that *There are 200 perfect logicians.*

This kind of indirect argument will appear often in the succeeding chapters. The readers should familiarize themselves with it.

One fun example of lingual self-referential behavior is the following story that in the beginning and in the end refers to itself.

A SELF-RECURRING STORY 1.5.16 There once was a girl who liked to travel from town to town, telling this story about herself. One day, while traveling in the dense forest, she entered a small village in a small clearing. She told them that she was hungry and tired, and then asked if she could exchange a telling of her story for some food and a place to sleep. But the villagers knew that only evil came from the dense forest, so they threw garbage at her, and chased her in large numbers. She was so overcome by these people that she stumbled and fell into a great blazing oven just outside the village. There she went up in a black cloud of smoke. This is always how her story ended, though, with her death in a fiery place. It seems that the myth and the miss had this end in common.

Let us end this discussion with a different version of Epimenides.

EXAMPLE 1.5.17 Epimenides, a Cretan, steps into an Athenian party and states that "All Cretans are Truth Sayers. It's a religious thing. We speak only the Truth." A young female student in the room says "Hey, Epimenides. Tell this bartender that I'm old enough to drink." Epimenides cannot tell this to the bartender since he is a Truth Sayer. Not knowing what else to do, he hangs his head and walks out of the party. We respect Epimenides because he did not contradict his first statement by telling the bartender that the lass was old enough for alcohol.

1.6 Exercises

1. Prove that *All cats are bald* is False.

2. Prove that *All birds lack feathers* is False.

3. Prove that *All people are liars* is False. Use a counterexample and a proof by contradiction.

4. Prove that *All statements are False* is False. Use a counterexample and a proof by contradiction.

5. Prove that *Left alone things do not change* is False. Use a counterexample to show that there is something out there that changes when left alone.

6. Prove that *Math is finite* is False by finding a counterexample.

7. Prove that *Nothing is known* is False. Use a counterexample and a proof by contradiction.

8. Prove that *No opinion is valid* is False. Use a proof by contradiction.

9. This is a hard one. Find the logical state of *I am lying* when it exists outside of the Epimenides Paradox.

10. Refer to # 9. See Example 1.5.17 for the definition of *Truth Sayer* If you are a Truth Sayer then can you speak *I am lying*?

11. Let P be a statement that has an unnamed logical state S. Does P have logical state S in every conversation that contains it?

Chapter 2

Sets

In this book, we will count the number of elements or objects that have a certain given property. We will constantly be answering the question *Just how many elements are there like that?* We will only count finite collections, but these finite collections can be difficult to count the traditional way. For example, in how many ways can you arrange six people in a row? The answer is not easily obtained by listing all of the arrangements or by simply making each of the arrangements with six people. Thus, we will have to find a different way to count these arrangements. Here is another one. How many pizzas can you make with exactly five different toppings if there are no repeated toppings and if there are ten toppings to choose from? These counting problems can become much more difficult with just a few changes in the wording. For example, consider the following counting problem. How many pizzas can you make with at most five different toppings if there are no repeated toppings and if there are ten toppings to choose from? These are typical of the counting problems that are our goal.

2.1 Set Notation

The language of mathematics is based on a grammar called *Set Theory*. This grammar gives us a brief and simple way to write down some very complex mathematical ideas. A *set* is a collection A such that given an object x, either x is in A, or x is not in A, but not both. As long as we can tell that any object is in A or not in A then A is a set. The collections we consider in this book will always be sets. The objects in A are called *elements.*

$x \in A$ is read as x *is in* A, or as x *is an element of* A.

$x \notin A$ is read as x *is not in* A, or as x *is not an element of* A.

The elements of a set are sometimes listed between *set braces*, $\{\quad\}$. For example, the following sets are described by listing the elements in them:

$$\{1,2\}, \quad \{a,m,z\}, \quad \{\text{book1}, \text{book2}, \text{book3}, \text{book4}\}.$$

The first set has elements 1 and 2 but no other elements. Observe that book3 is an element in the third set but not an element of the second set. Also, Algebra is not an element of the third set unless it is book1, book2, book3, or book4.

An important set is called the *empty set.* This is the set that has no elements. In this book, the empty set is written in the following ways:

$\emptyset. \quad \{\,\}$

Thus, given an object x we know with certainty that x *is not in* $\emptyset$. In terms of our counting theme, the empty set is the set with no elements, or 0 elements. For example, the set of cards in a standard deck of 52 that are labeled by 11 make up the set $\{\}$. There are no such cards.

2.2 Predicates

The problems we are working on take place in a finite set called a *universal set* or a *universe.* This universe changes from problem to problem. For instance, if we are counting the number of red, white, and blue flags in the world, then this problem takes place in a universe of flags of the world. This set is finite. There are only a finite number of flags in this world. The set of all numbers that appear in a standard deck of 52 cards is finite. In fact, we can list its nine elements as $\{2, 3, 4, 5, 6, 7, 8, 9, 10\}$.

Let us keep in mind that this universe that we are defining is a finite set and that it changes from problem to problem. These universes will never be the universe we live in since our surroundings are infinite once you count the abstract ideas that make up our thoughts.

The most effective means of listing a set is called a *predicate*. A predicate is a partial sentence that describes objects.

EXAMPLE 2.2.1 1. *red, white, and blue flags* is a predicate describing flags.

2. *Positive even whole numbers less than 12* is a predicate describing the set $\{2, 4, 6, 8, 10\}$.

3. The predicate $x^2 - 4 = 0$ can be used to describe the set $\{-2, 2\}$ as follows.

$$\{\text{real numbers } x \mid x^2 - 4 = 0\} = \{-2, 2\}$$

Read the symbol $\mid$ as *such that* or as *with the property that*. Then $x^2 - 4 = 0$ is a perfectly good predicate to describe the objects in a set.

4. Consider $\{x \mid x \text{ is a person on Earth}\}$. The predicate is x *is a person on Earth.* This is a set that in this computer age could be given as a finite complete list of people on the Earth, but which should not be given as that large list in this book.

5. There is also the predicate *is a book*, which describes $\{x \mid x \text{ is a book}\}$.

Let A be a set in a universal set $\mathcal{U}$. The *complement of* A is the set

$$A' = \{x \in \mathcal{U} \mid x \notin A\}.$$

This is the set of elements that are not in A. This is what is outside of A in $\mathcal{U}$. Think of A as a circle, and then A' is what is outside that circle.

EXAMPLE 2.2.2 Let $\mathcal{U} = \{0, 1, 2, 3, 4, 5, 6, 7, 8, 9, 10\}$.

1. Let $A = \{0, 1, 2, 3, 4, 5\}$. Then $A' = \{6, 7, 8, 9, 10\}$. These are the numbers in $\mathcal{U}$ that are not in A.

2. Let $B = \{1, 3, 5, 7, 9\}$. Then $B' = \{0, 2, 4, 6, 8, 10\}$, the numbers not in B.

3. Let C be the set of even numbers in $\mathcal{U}$. Then $C' = \{1, 3, 5, 7, 9\}$ which is the set of odd numbers in $\mathcal{U}$.

4. Let $\mathcal{U}$ be the set of children of age at most 6 years, and let $B =$ the set of boys of age at most 6 years. Then B' is the set of children of age at most 6 years that are not boys. These are the girls of age at most 6 years.

There are two complements that should be remembered. They come up often. We will assume the existence of a finite universal set $\mathcal{U}$ in all that follows. Since $\mathcal{U}$ contains everything being considered, no x satisfies $x \notin \mathcal{U}$. Thus, the complement of $\mathcal{U}$ is empty.

$$\mathcal{U}' = \emptyset$$

Since nothing is in $\emptyset$, $x \notin \emptyset$ for each $x \in \mathcal{U}$. Hence, the complement of $\emptyset$ is $\mathcal{U}$. In other words,

$$\emptyset' = \mathcal{U}.$$

2.3 Subsets

When one set is contained in another set we say that the smaller is a *subset* of the larger. This is an intuitive way of thinking of subsets, but we will need a more precise definition of subset. Let A and B be sets. We say that B *is a subset of* A, and we write $B \subset A$, if

Given $x \in B$ then $x \in A$.

When paraphrased, we say that $B \subset A$ if *each element of* B *is an element of* A. One more way of saying $B \subset A$ is that *if we are given an* $x \in B$ *then we can show that* $x \in A$.

EXAMPLE 2.3.1 Let $\mathcal{U}$ = the set of whole numbers between 0 and 25 inclusive.

1. $\{1,2,3\} \subset \{1,2,3,4,5\}$. This is True because evidently 1, 2, and 3 are elements of $\{1,2,3,4,5\}$.

2. $\{1,3\} \subset \{x \in \mathcal{U} \mid x \text{ is an odd number}\}$. This is True because evidently 1 and 3 are odd numbers in $\mathcal{U}$.

3. Let B = the set of positive whole numbers between 1 and 25 inclusive. Each element of B is a whole number between 0 and 25 inclusive. Hence $B \subset \mathcal{U}$.

There are two examples of subsets that we should always be aware of.

Let A be a set. Evidently each element of A is an element of A. This is just mathematical double talk. Thus we conclude that

$A \subset A$.

Now consider the other end of the problem. Let A be a set. We will argue that

$\emptyset \subset A$.

To see this, suppose to the contrary that $\emptyset \not\subset A$. This will lead us to a False statement. By the definition of subset, there is an element of $\emptyset$ that is not in A. But that means that $\emptyset$ has an element, which is not True of the empty set. This mathematical mistake, called a *contradiction*, shows us that we began with a Falsehood. You see, if we make no mistakes then a Truth leads to a Truth. The assumption that $\emptyset \not\subset A$ leads us to the untruth that $\emptyset$ contains an element. Thus it must be that we did proceed from a False premise. Hence $\emptyset \subset A$.

We will have need of a list of all subsets of a set. Some counting problems use them all. Let A be a set. The set of all subsets of A is called the *power set of* A and it is denoted by

$$\mathcal{P}(A) = \text{the set of all subsets of } A.$$

EXAMPLE 2.3.2 Let $A = \{\} = \emptyset$. Then A has exactly 0 elements, so that any subset of A has exactly 0 elements. The only set with 0 elements is $\emptyset$. Then $\emptyset$ is the only subset of A. So $\mathcal{P}(A) = \{\emptyset\}$.

EXAMPLE 2.3.3 Let $A = \{a\}$. Then A has exactly one element so that every subset of A will have at most 1 element. That is, they will have exactly 1 element or no elements. These subsets are $\{\}$ which has no elements, and $\{a\}$ which has exactly 1 element. So $\mathcal{P}(A) = \{\emptyset, \{a\}\}$.

EXAMPLE 2.3.4 $A = \{a, b\}$. Then A has exactly 2 elements. Its subsets will then have exactly 2, or exactly 1, or no elements at all. The subset with no elements is $\{\}$, the subsets with exactly 1 element are $\{a\}, \{b\}$, and the only subset with exactly 2 elements is $\{a, b\}$. So $\mathcal{P}(A) = \{\emptyset, \{a\}, \{b\}, \{a, b\}\}$.

2.4 Union and Intersection

In this section we study several important operations on sets. The *union* of sets A and B is

$$A \cup B = \{x \mid x \in A \text{ or } x \in B\}.$$

In using *or* to define $A \cup B$ we use the *inclusive or*. This *or* allows the possibilities that $x \in A$, that $x \in B$, and that x is in both A and B.

The *intersection* of A and B is

$$A \cap B = \{x \mid x \in A \text{ and } x \in B\}.$$

In order that $x \in A \cap B$ it must be True that x is in both A and B.

EXAMPLE 2.4.1 Let $A = \{a, b, c, d\}$ and $B = \{c, d, e\}$. Then

1. $A \cup B = \{a, b, c, d, e\}$ because these are the elements in either A or B.

2. $A \cap B = \{c, d\}$ because these are the elements in both A and B.

EXAMPLE 2.4.2 Let $A = \{1, 2, 3, 4, 5\}$ and $B = \{-2, -1, 0, 1, 2, 3\}$. Then

1. $A \cup B = \{-2, -1, 0, 1, 2, 3, 4, 5\}$ because these are the elements is either A or B.

2. $A \cap B = \{1, 2, 3\}$ because these are the elements in both A and B.

There are several rules of manipulation involving $\cup$, $\cap$, and complement $()'$. The first are the distributive laws. Let A, B, and C be sets.

$$A \cap (B \cup C) = (A \cap B) \cup (A \cap C)$$

and

$$A \cup (B \cap C) = (A \cup B) \cap (A \cup C)$$

These distributive laws are True because the use of *and* and *or* in the language satisfies a distributive law of sorts. For instance, the sentences *P, and Q or R* has the same meaning as the sentence *P and Q, or P and R.*

Then there are the important *DeMorgan's laws.* Let A and B be sets in a universe $\mathcal{U}$. Then

$$(A \cap B)' = A' \cup B'$$

and

$$(A \cup B)' = A' \cap B'.$$

Note the change in the first law from $\cap$ to $\cup$ after we distribute the complement symbol. Note the change in the second law from $\cup$ to $\cap$ after we distribute the complement symbol. These laws allow us to replace *neither P nor Q* with *not P and not Q.*

EXAMPLE 2.4.3 DeMorgan's laws show us that these statements mean the same thing.

1. It is neither Tuesday nor the second.

2. It is not Tuesday and it is not the second.

Note the change from *or* to *and* after we distribute the negation.

EXAMPLE 2.4.4 Let $A = \{2,3,4\}$, $B = \{3,4,5\}$, and let $\mathcal{U} = \{1,2,3,4,5,6,7,8,9\}$. Then

$$(A \cup B)' = \{2,3,4,5\}' = \{1,6,7,8,9\}$$

while

$$A' \cap B' = \{1,5,6,7,8,9\} \cap \{1,2,6,7,8,9\} = \{1,6,7,8,9\}.$$

We observe that $(A \cup B)' = A' \cap B'$, just as DeMorgan's law predicts.

2.5 Exercises

1. Describe these sets with predicates.

    1. $\{x \mid 0 \leq x \leq 1\}$ Answer: $\{x \mid x$ is a number between 0 and 1 inclusive$\}$.

    2. $\{red, yellow, orange, blue, violet, green, brown, black\}$ Answer: $\{x \mid x$ is one of the primary colors$\}$.

2. Make a list of the elements in the sets described by these predicates.

    1. The whole numbers between 0 and 10 inclusive.

    2. The names of the people in your family.

    3. The colors of crayons in a small box of crayons

3. Why is the set of children on Earth a subset of the set of people aged at most 21? Answer: If x is a child then the age of x is less than 21. Thus $\{x \mid x$ is a child$\} \subset \{x \mid x$ is at most 21 years old$\}$.

4. Write down $\mathcal{P}(\{1, 2\})$.

5. Write down $\mathcal{P}(\{a, b, c\})$. Answer: $\mathcal{P}(\{a, b, c\})$ is made up of subsets with no elements, 1 element, 2 elements, or 3 elements. There are 2^3 such elements. Thus $\mathcal{P}(\{a, b, c\}) = \{\emptyset, \{a\}, \{b\}, \{c\}, \{a, b\}, \{a, c\}, \{b, c\}, \{a, b, c\}\}$.

6. Write down $\mathcal{P}(\{a, b, c, d\})$.

7. Let $A = \{2, 3, 4, 5\}$, $B = \{2, 3, 4, 5, 6\}$, $C = \{1, 2, 6\}$ and let $\mathcal{U} = \{0, 1, 2, 3, 4, 5, 6, 7, 8, 9\}$. Find and compare these sets.

    1. $A \cap (B \cup C)$ and $(A \cap B) \cup (A \cap C)$. Answer: $A \cap (B \cup C) = \{2, 3, 4, 5\} \cap \{1, 2, 3, 4, 5, 6\} = \{2, 3, 4, 5\}$.

    2. $A' \cap B'$ and $(A \cup B)'$. Answer: $A' \cap B' = \{0, 1, 7, 8, 9\}$. $(A \cup B)' = \{0, 1, 7, 8, 9\}$. Notice that $A' \cap B' = (A \cup B)'$.

    3. $A' \cup B'$ and $(A \cap B)'$.

8. Show that the following are equal.

1. $(A \cap B \cap C)' = A' \cup B' \cup C'$.
2. $(A \cup B \cup C)' = A' \cap B' \cap C'$.

Chapter 3

Venn Diagrams

Survey problems present a difficulty in reading the survey results accurately and properly. For example, if we know that 35 people read the *New York Times* and that 25 people read the *Washington Post*, then how many read both papers? We can answer this problem if we know a little bit more about the survey, like how many were surveyed, and if we know our first principle of counting. We can use this principle to dissect even more complex surveys.

3.1 Inclusion-Exclusion Principle

The number of elements in a set is denoted by

$$n(A) = \text{ the number of elements in } A.$$

Suppose that you have sets A and B. To count the number of elements in $A \cup B$ you might first try adding $n(A)$ and $n(B)$. But when you do this you are counting the elements in $A \cap B$ twice: once for A and once for B. This is the only overlap between A and B so we have found that

$$n(A \cup B) = n(A) + n(B) - n(A \cap B).$$

We subtract the number of elements $n(A \cap B)$ because we have counted it twice before when we calculate $n(A)$ and $n(B)$. The subtraction compensates for the double counting. This formula is called the *Inclusion/Exclusion Principle.* Notice that $n(A \cup B)$ is not simply $n(A) + n(B)$. The subtraction gives us a balanced equation.

EXAMPLE 3.1.1 Suppose that A has 25 elements, that B has 20 elements, and that $A \cap B$ has 10 elements. Then we can calculate the number of elements in $A \cup B$.

$$\begin{aligned} n(A \cup B) &= n(A) + n(B) - n(A \cap B) \\ &= 25 + 20 - 10 \\ &= 35. \end{aligned}$$

EXAMPLE 3.1.2 Suppose that A has 50 elements, that B has 30 elements and that $A \cup B$ has 60 elements. Then we can calculate the number of elements in $A \cap B$. Since $n(A \cup B) = n(A) + n(B) - n(A \cap B)$ we have

$$\begin{aligned} 60 &= 50 + 30 - n(A \cap B) \\ n(A \cap B) &= 50 + 30 - 60 \\ n(A \cap B) &= 20 \end{aligned}$$

There are 20 elements in the intersection of A and B.

Two sets A and B are *disjoint* if $A \cap B = \emptyset$, or equivalently if A and B do not contain common elements. In the event that A and B are *disjoint* then the inclusion/exclusion principle simplifies somewhat. Suppose that A and B are sets and that A and B are disjoint. Then $n(A \cap B) = n(\emptyset) = 0$. Hence $n(A \cup B) = n(A) + n(B) - n(A \cap B) = n(A) + n(B)$. That is,

$$n(A \cup B) = n(A) + n(B) \text{ if } A \text{ and } B \text{ are disjoint.}$$

EXAMPLE 3.1.3 Suppose that 50 people voted for X for president and 35 voted for Y for president. Assuming that a voter cannot vote for both candidates, how many people voted overall? To answer this question, let A be the set of those who voted for X and let B be the set of those who voted for Y. Then our assumption implies that A and B are disjoint, that $A \cap B = \emptyset$. The people who voted are in the set $A \cup B$. Hence, for this problem,

$$\begin{aligned} n(A \cup B) &= n(A) + n(B) \\ n(A \cup B) &= 50 + 35 = 85. \end{aligned}$$

Thus, 85 people voted overall.

It is interesting to note that there is an inclusion/exclusion principle for the union of three sets. It is found as follows. Let A, B, and C be sets. We wish to find $n(A \cup B \cup C)$, the number of elements in the three sets. First let us consider

$$n(A) + n(B) + n(C).$$

This sum is the number we are interested in because to find it we are counting the overlaps in the sets twice. That is, we are counting $A \cap B$ once when we count A and once when we count B. We are counting $A \cap C$ once when we count A and once when we count C. We are counting $B \cap C$ once when we count B and once when we count C. Since in the final analysis we will count elements only once, we will subtract the doubly counted regions from $n(A) + n(B) + n(C)$. This yields

$$n(A) + n(B) + n(C) - n(A \cap B) - n(A \cap C) - n(B \cap C). \quad (3.1)$$

There is yet another intersection $A \cap B \cap C$ to be considered. The intersection $A \cap B \cap C$ is counted once each time we counted $n(A)$, $n(B)$, and $n(C)$. It is then removed from the count each time we counted $n(A \cap B)$, $n(A \cap C)$, and $n(B \cap C)$. Thus, $n(A \cap B \cap C)$ has been added into the expression (3.1) three times and then subtracted three more times. The result is that (3.1) does not count $n(A \cap B \cap C)$. To compensate we add it in one more time.

$$\begin{aligned} n(A \cup B \cup C) = n(A) &+ n(B) + n(C) \\ &- n(A \cap B) - n(A \cap C) - n(B \cap C) \\ &+ n(A \cap B \cap C) \end{aligned}$$

This is the *inclusion/exclusion principle* for three sets.

3.2 Two-Circle Venn Diagrams

The Inclusion/Exclusion Formulas in the previous section are useful for some analysis, but in some cases they fall short.

EXAMPLE 3.2.1 Suppose that we poll 50 people, 25 of whom watch TV only, and 15 of whom listen to the radio only. All those polled fall into one category or the other. How many watch TV and listen to the radio?

There seems to be two sets involved in this problem. Let $T =$ the set of people polled who watch TV and let $R =$ be the set of people polled who listen to the radio. We are asked to find the number $n(T \cap R)$. We know that there are 25 people polled who watch TV only, but this is not all of T. It is only a portion of it. We know that 15 people polled listen to the radio only, but this is not all of R. Thus we do not know the numbers $n(T)$ and $n(R)$, so we cannot use the inclusion/exclusion principle $n(T \cup R) = n(T) + n(R) - n(T \cap R)$.

Instead, we argue as follows. The number polled is 50, so $n(T \cup R) = 50$. Anyone polled will either watch TV only, or they will listen to radio only, or they will watch TV and listen to the radio. The number watching TV and listening to radio is $n(T \cap R)$. Thus,

$$\begin{aligned} n(T \cup R) &= n(\text{watch TV only}) + n(\text{listen to radio only}) + n(T \cap R) \\ 50 &= 25 + 15 + n(T \cap R) \\ 10 &= n(T \cap R). \end{aligned}$$

It follows that 5 watch TV and listen to the radio.

The above problem would be simpler if there was a more visual expression of the data. Below is a visual aid called a *Venn diagram* which visually represents the problem about the TV and that radio that we are trying to solve. The diagram contains much more information than we need to solve the problem, but that is fine. It will be a useful tool in solving many more problems than this one.

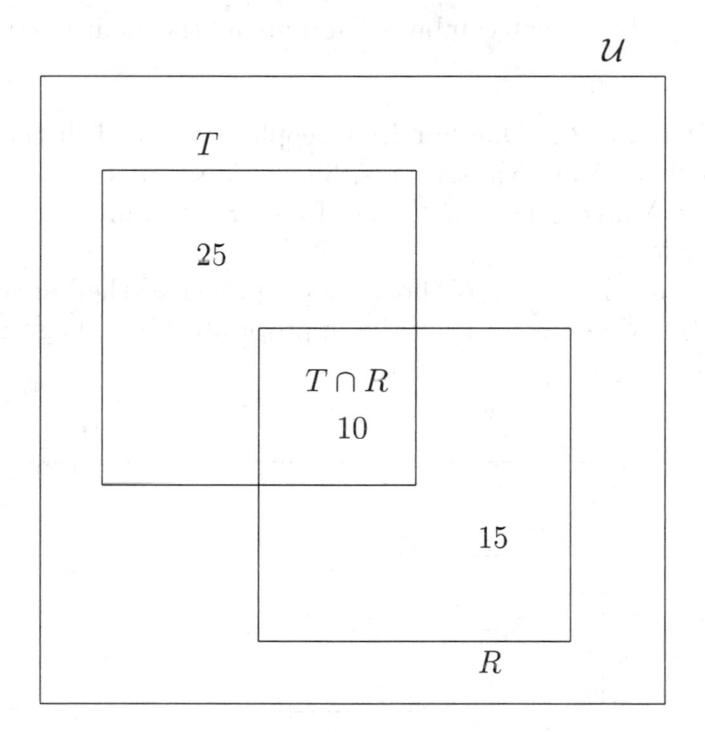

The Venn diagram associated with the TV and radio problem is the one above.

The number 25 represents the fact that 25 watch TV only. The 25 does not imply that $n(T) = 25$. No indeed. The number 25 only represents the number of elements in the region in the square T that does not contain $T \cap R$. These are the people who watch TV and do not listen to radio. The region in square R labeled with 15 represents those polled who only listen to the radio. There are 15 such people, thus the number 15 in the region. The number 15 labels the region in R that does not contain $T \cap R$. We read in the problem that the number in $T \cup R$ is 50, so the sum of the three smaller regions must be 50. In formulas

$$n(T \cup R) = 50 = 25 + 15 + n(T \cap R)$$

so that $n(T \cap R) = 50 - 25 - 15 = 10$. Then 10 labels the $T \cap R$ region. See the diagram. This is the same value we got above. In the future I will not need to include so much detail to solve the problem.

Some problems will combine diagram and the inclusion/exclusion principle.

EXAMPLE 3.2.2 One hundred people are polled in the Bronx. We find that 75 are Yankee fans, 35 are Mets fans, and 10 people polled are Yankees fans and Mets fans. How many were Yankees and Mets fans?

There are three sets: $\mathcal{U}$ those people polled, Y the Yankees fans, and M the Mets fans. Draw the appropriate Venn diagram as we did below.

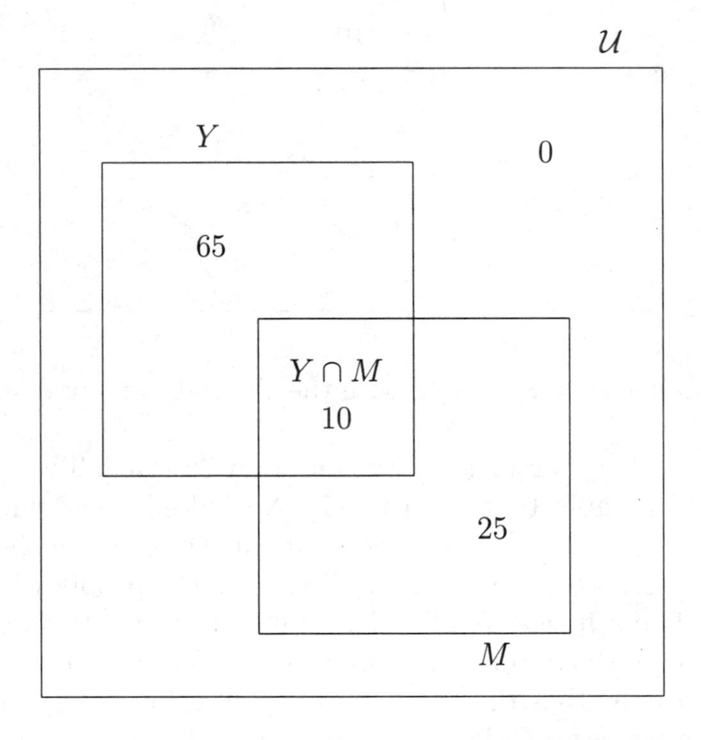

The region corresponding to those people that are Yankees fans and Mets fans is $Y \cap M$, and there are 10 people in this region. Fill in the Venn diagram. $n((Y \cap M)) = 10$ so the region inside the two squares is labeled 10. There are a total of $n(Y) = 75$ Yankees fans and we have already accounted for 10 of them so the region $Y \cap M'$, which is the region in Y that does not include $Y \cap M$, is labeled with $75 - 10 = 65$ people. Notice that the sum of the numbers in the regions contained in Y is 75. The region M contains a total of $n(M) = 35$ Mets fans, and we have already accounted for 10 of them in $Y \cap M$. Thus, the region $Y' \cap M$, which is the region of Mets fans that are not Yankees fans, has exactly $35 - 10 = 25$ people in it. There is one more region to label. It is the region corresponding to $(Y \cup M)'$. This is the region outside of $Y \cup M$. This is the region outside the squares Y and M.

To find the number of people in this region, we find

$$n(Y \cup M)' = n(\mathcal{U}) - n(Y \cup M).$$

To find $n(Y \cup M)$ we add up the regions contained in $Y \cup M$. That sum is

$$n(Y \cup M) = 65 + 25 + 10 = 100.$$

Since $n(\mathcal{U}) = 100$, we see that $n(Y \cup M)' = n(\mathcal{U}) - n(Y \cup M) = 0$. Those surveyed were either Yankees fans or Mets fans. I have labeled the Venn diagram to reflect these values.

Here is an interesting problem that is most efficiently done by labeling a Venn diagram.

EXAMPLE 3.2.3 Thirty people polled stated that they liked water and not milk, 45 people polled stated that they liked milk, and 25 stated that they did not like either drink. How many people were polled?

Let W denote the set of people who liked water, let M denote the set of people who liked milk, and box it all in with a rectangle labeled $\mathcal{U}$. We want to know $n(\mathcal{U})$.

The region labeled by 30 represents the people polled who liked water only. It does not include the people in the intersection $W \cap M$. The square M is labeled by 45. I have placed the label 45 in the center so that you will know it represents something different from the other labels. Unlike the number label 30, the number label 45 accounts for all of the square M. Those not liking either drink would be people outside of the two squares. Thus, $(W \cup M)'$, the region outside of the squares, is labeled by 25.

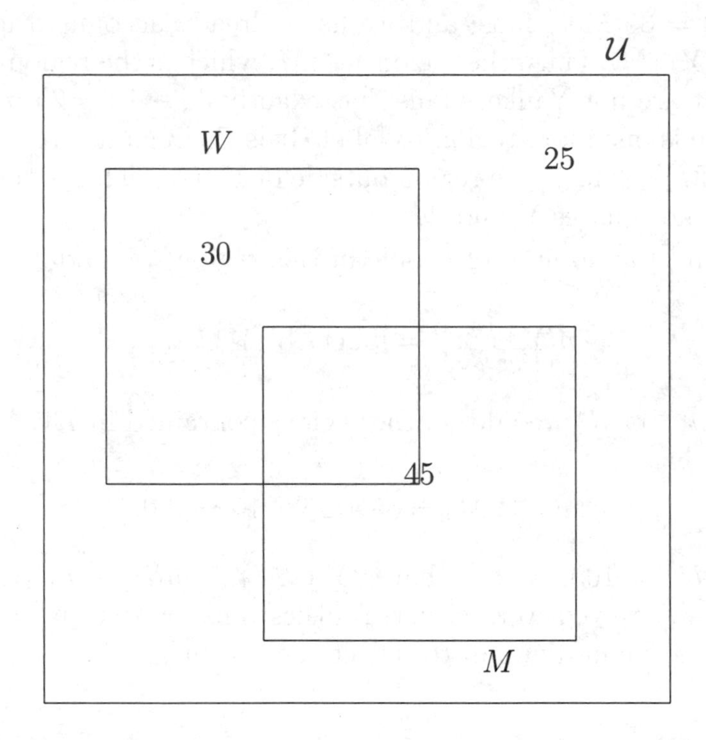

The number polled will be the sum of the labels of each region.

$$n(\mathcal{U}) = 30 + 45 + 25 = 100.$$

One hundred people were polled.

3.3 Three-Square Venn Diagrams

Two-square Venn diagrams can be used when there are two parameters to the survey being discussed. But if we include three parameters, the two-square Venn diagram must be replaced by a three-square Venn diagram, like the one above. The capital letters reside next to squares that they label. The lowercase letters are the numbers of elements in the connected regions of the diagram in which they reside. For example, g counts the number of elements in the region that is inside each square, while a counts the number of elements in the region that is in A only. h counts the number of elements in the region that is outside of each of the three squares.

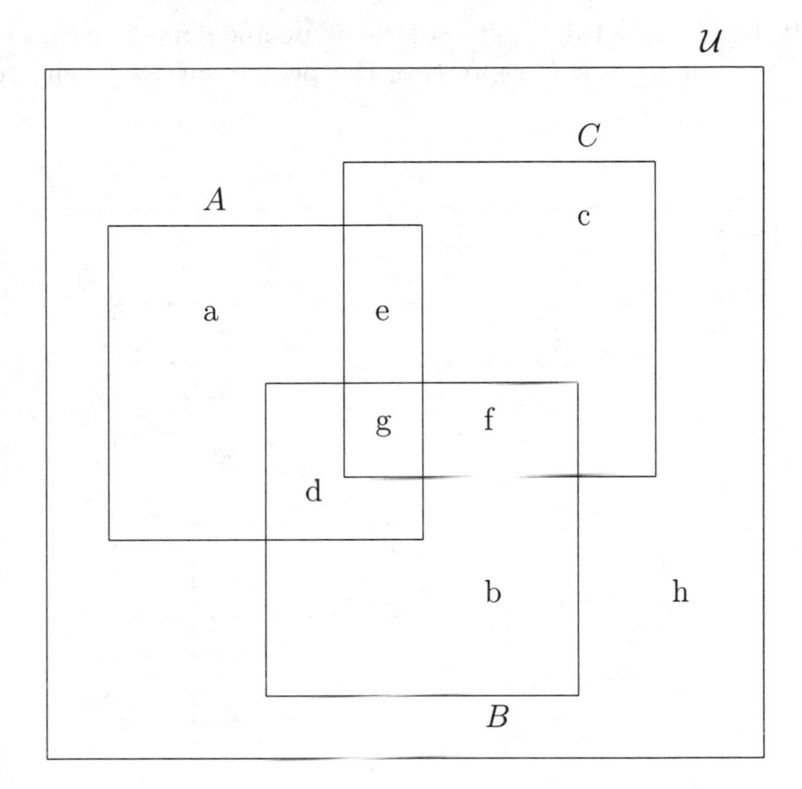

Larger Venn diagrams prove to be impractical, so this is the limit of our Venn diagram discussion. Let us see how some problems with three parameters are solved using a three-square Venn diagram.

EXAMPLE 3.3.1 Four hundred people are surveyed, and each responds to the survey. One hundred and fifty read the *Times*, 125 read the *News*, and 100 read the *Post.* Seventy five read the *Times* and the *News*, 50 read the *Times* and the *Post*, and 25 read the *News* and the *Post.* Five read all three newspapers. Draw and label the associated Venn diagram.

We begin by listing the information in the problem.

$$\begin{array}{lll} n(T) = 150 & n(T \cap N) = 75 & \\ n(N) = 125 & n(T \cap P) = 50 & \\ n(P) = 100 & n(N \cap P) = 25 & n(T \cap N \cap P) = 5 \end{array}$$

Draw the three-square Venn diagram (A) pictured below. The square T represents the set of people surveyed who read the *Times*, the square labeled N represents the people surveyed who read the *News*, the square P represents the people surveyed who read the *Post*.

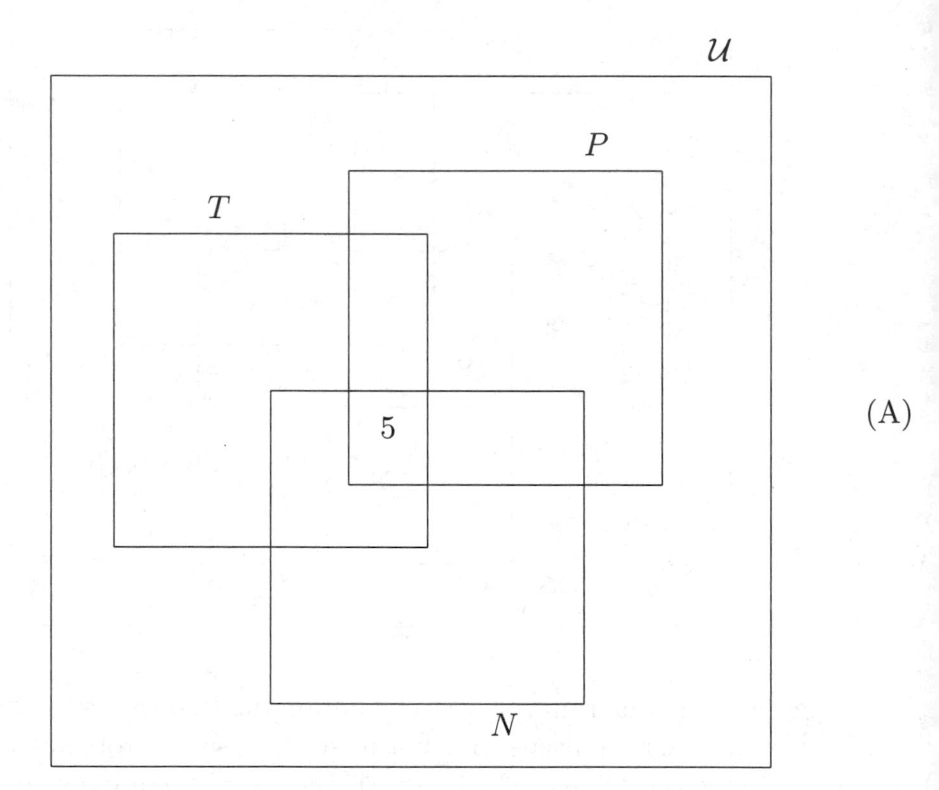

(A)

To fill in the disjoint regions of the diagram, begin with the smallest region and work your way out. We are told that 5 read all three newspapers, so $n(T \cap N \cap P) = 5$. Fill in the small central region, the one that corresponds to $T \cap N \cap P$, with a 5, as in (A).

Next consider the regions corresponding to the intersections of two sets. Since $n(N \cap P) = 25$ and since we already have 5 elements of $N \cap P$, the regions in $N \cap P$ are labeled 5 and 20, as in diagram (B).

Because $n(T \cap P) = 50$ and since we already have 5 elements of $T \cap P$, the regions in $T \cap P$ are labeled 5 and 45. See diagram (B). Since $n(T \cap N) = 75$ and since we already have 5 elements of $T \cap N$, the regions in $T \cap N$ are labeled as 5 and 70.

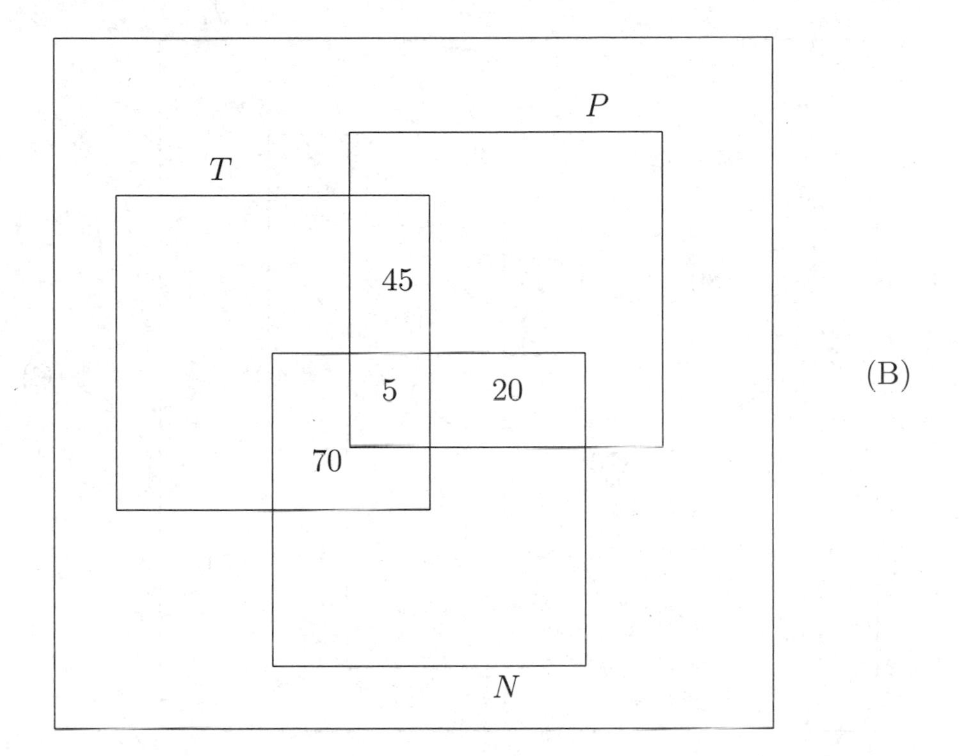

(B)

To fill in the unlabeled regions in (B), note that $n(T) = 150$ and that we have identified $70 + 45 + 5 = 120$ people in T. Thus, $150 - 120 = 30$ labels the region in T, as in the diagram below. Since $n(N) = 125$ and since we have identified $70 + 20 + 5 = 95$ people in N, we label the remaining region of N with $125 - 95 = 30$. Since $n(P) = 100$ and since we have identified $45 + 20 + 5 = 70$ people in P, we label the remaining region in P with $100 - 70 = 30$. See diagram (C).

The region outside of the three squares must be filled in. This is found by taking the total number surveyed, 400, and subtracting the sum of the values

$$30 + 30 + 30 + 70 + 45 + 20 + 5 = 230$$

in the three squares. Then $400 - 230 = 170$ will label the region outside the squares, as in (C).

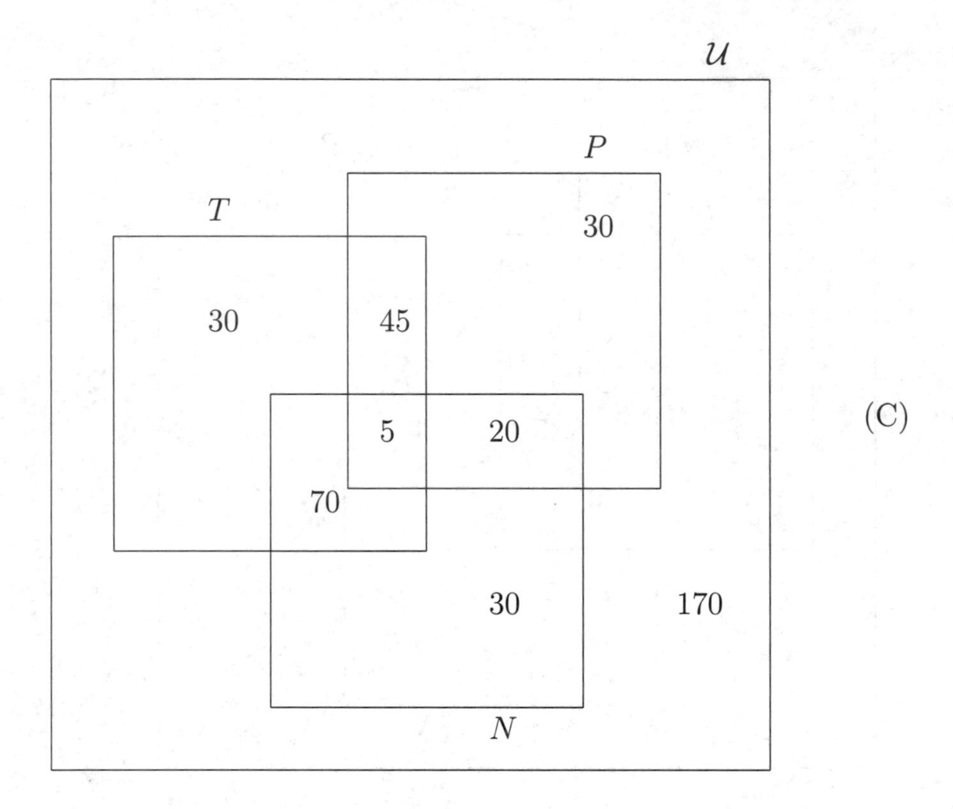

EXAMPLE 3.3.2 Consider the Venn diagram (C) that was labeled with the data in the above problem. Use diagram (C) to answer the following questions.

1. How many surveyed did not read the *Times*, the *News*, or the Post?

2. How many read the *Post* but neither the *News* nor the *Post*?

3. How many read exactly one newspaper?

4. How many read exactly two newspapers?

1. The region corresponding to the people who did not read the *Times* or the *News* or the *Post* is the region outside the three squares. The number there is 170.

2. The region in T and not in N or P is the region in T that does not overlap N and that does not overlap P. This region is labeled by 30.

3. The regions corresponding to people who read exactly one newspaper are those labeled by 30. Those regions sit inside one and only one square, so these regions represent people who read one and only one newspaper. There are a total of $30 + 30 + 30 = 90$ people who read exactly one newspaper.

4. The regions corresponding to people who read exactly two newspapers are those labeled by 70, 45, and 20. These are regions that are in two squares but not in all three squares. There are a total of $70 + 45 + 20 = 135$ people who read two and only two newspapers.

In some problems involving Venn diagrams there is the need for some algebra. Here is an example.

EXAMPLE 3.3.3 Five hundred people are surveyed and each responds. There are three banks in town called bank A, bank B, and bank C. Fifty people do not have money in any of the banks, while Ten people have money in each bank. The number of people who have money in banks A and B is also the number of people who have money in banks B and C, which is the number of people who have money in banks A and C. There are 150 people who have money in bank A, there are 125 people who have money in bank B, and there are 100 people who have money in bank C. Draw and label the associated Venn diagram.

We begin by listing the information in the problem. We need a little algebra here. The regions that are in two squares each have an equal but unknown number of elements. Call that value x and list your data.

$$\begin{array}{lll} n(A) = 150 & n(A \cap B) = x & \\ n(B) = 125 & n(A \cap C) = x & \\ n(C) = 100 & n(B \cap C) = x & n(A \cap B \cap C) = 10 \end{array}$$

Make the three square Venn diagram below. The square A represents the set of people with money in bank A, the square labeled B represents the people surveyed who have money in bank B, and the square C represents the people surveyed who have money in bank C.

Since the squares represent people who have money in at least one bank, the region outside of the squares represents the people who do not have money in any bank. There are 50 people not in the squares. 10 have money in all three banks, so 10 labels the smallest region.

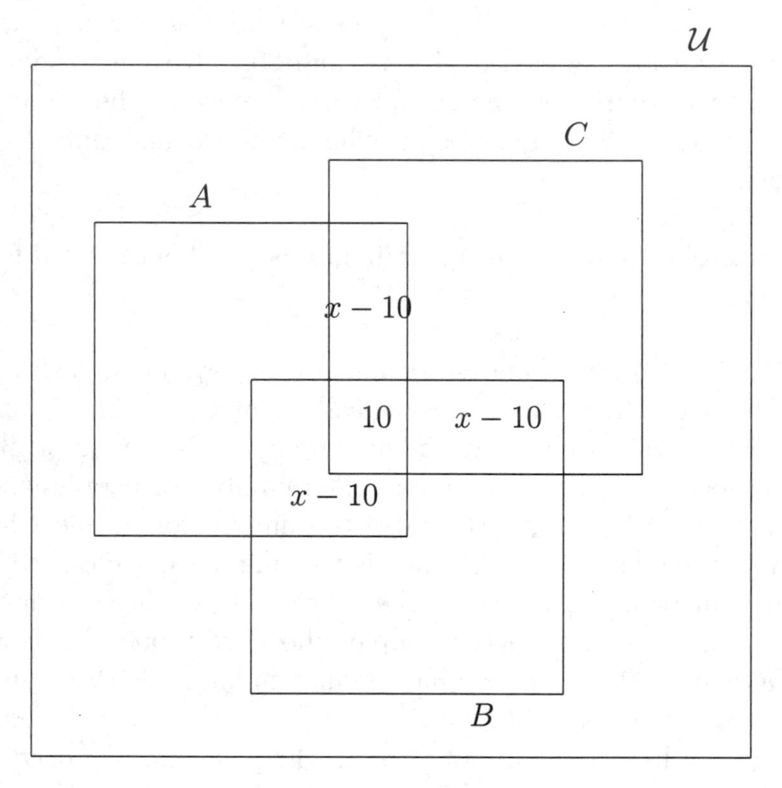

Since $n(A \cap B) = n(A \cap C) = n(B \cap C) = x$ and since we already have 10 elements of these intersections, each intersection is labeled by 10 and $x-10$. Note that the sum of the labels in the intersections is $10 + (x - 10) = x =$ the number of people in the intersections of two squares.

The remaining portions of the circular regions are labeled as follows. Since $n(A) = 150$, in square A, the unlabeled region represents $150-(x+(x-10)) = 160-2x$ people. Since $n(B) = 125$, in square B, the unlabeled region represents $125 - (x + (x - 10)) = 135 - 2x$ people. Since $n(C) = 100$, in square C, the unlabeled region represents $100 - (x + (x - 10)) = 110 - 2x$ people.

The region $(A \cup B)'$ of the Venn diagram corresponds to people who do not have money in any of the three banks. The number given to us by the problem is $n(A \cup B)' = 50$.

We have constructed the Venn diagram with labels below.

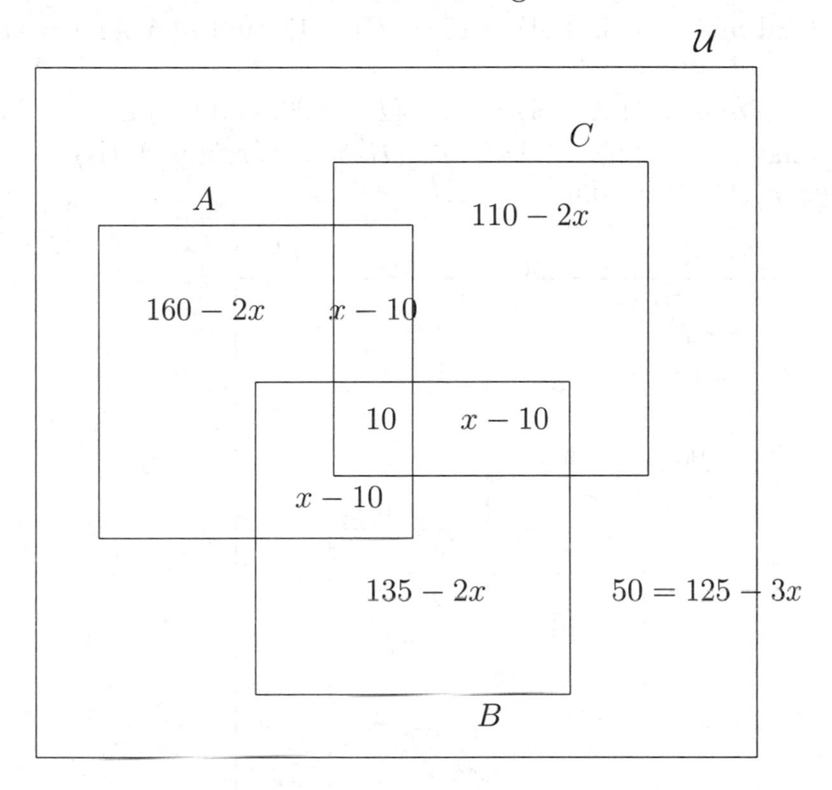

EXAMPLE 3.3.4 Refer to the problem above. Find the number of people with money in a pair of banks.

We are looking for the number x in the above Venn diagram. Calculate the label for the region outside of the three squares.

$$500 - (160 - 2x + 135 - 2x + 100 - 2x + x - 10 + x - 10 \\ +x - 10 + 10) = 500 - (375 - 3x) = 125 + 3x$$

The given data also shows us that this region must also be labeled by 50. So $50 = 125 - 3x$ labels the region outside of the squares. Solving for x we find that $50 = 125 - 3x$ implies that $3x = 75$, and so $x = 25$.

3.4 Exercises

In each of these problems draw the appropriate Venn diagram.

1. Find $n(A \cap B)$ if $n(A) = 10$, $n(B) = 15$, and $n(A \cup B) = 17$. Answer: $n(A \cap B) = 8$.

2. Find $n(A \cap B)$ if $n(A) = 15$, $n(B) = 20$, and $n(A \cup B) = 35$.

3. Find $n(A \cup B)$ if $n(A) = 100$, $n(B) = 105$, and $n(A \cap B) = 10$. Answer: $n(A \cup B) = 195$

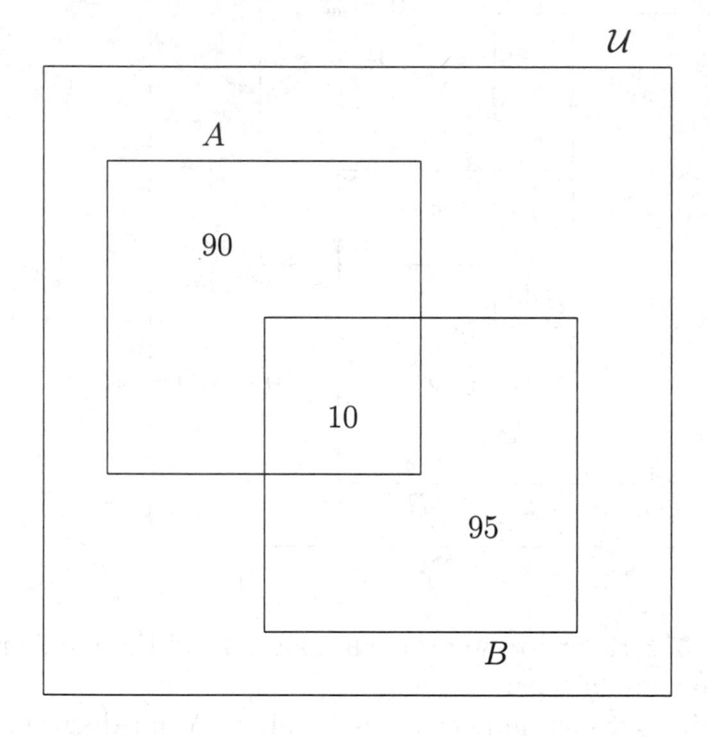

4. Find $n(A \cup B)$ if $n(A) = 9$, $n(B) = 11$, and if A and B are disjoint. Answer: $n(A \cup B) = 20$.

5. A poll of 150 people was taken about questions A and B on a test. One hundred said they answered A correctly, 100 said they answered B correctly, 75 said they answered both A and B correctly. How many answered neither problem correctly? Answer: 25.

6. One hundred people were polled about their dorm rooms. Seventy five answered that they had heat in their room, 50 answered that they had a rug in their room, and 25 people had neither heat nor a rug. How many polled answered that they had heat and a rug

in their room? Answer: 50.

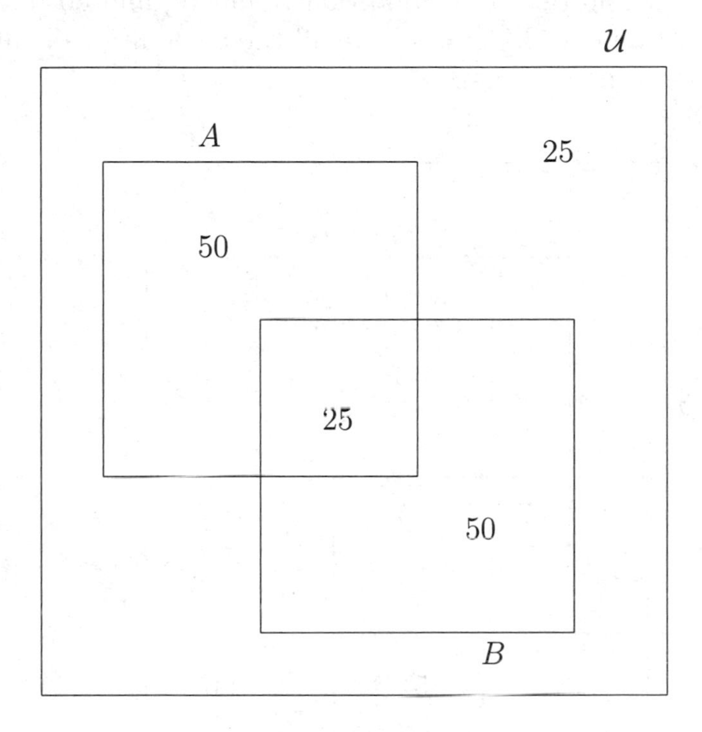

7. A different kind of multiple choice exam was taken by 150 students. The exam had one question whose answers were labeled a and b. You could answer the question by doing one of the following. You could (i) shade neither answer, (ii) shade-in a, (iii) shade-in b, or (iv) shade-in both a and b. Seventy five people taking the exam answered a to the question, 75 answered b to the question, and 25 did not shade-in a and did not shade-in b. Using only two squares in a Venn diagram, find the number of people who shaded in a and b. Answer: 25

8. One thousand horticulturists grow red, yellow, and white roses. Seven hundred grow red roses, 500 grow yellow roses, and 300 grow white roses. Of these horticulturists, 250 grow red roses and yellow roses, 200 grow red roses and white roses, 100 grow yellow roses and white roses, and 50 grow all three color roses. Fill in the appropriate Venn diagram.

9. Two hundred and fifty rats are evaluated for conditions L, M, and N. It is found that 100 rats have condition L, 80 rats

have condition M, and 75 rats have condition N. Fifty rats have conditions L and M, 40 rats have condition L and N, and 30 rats have conditions M and N. Five rats have all three conditions. Fill in the appropriate Venn diagram.

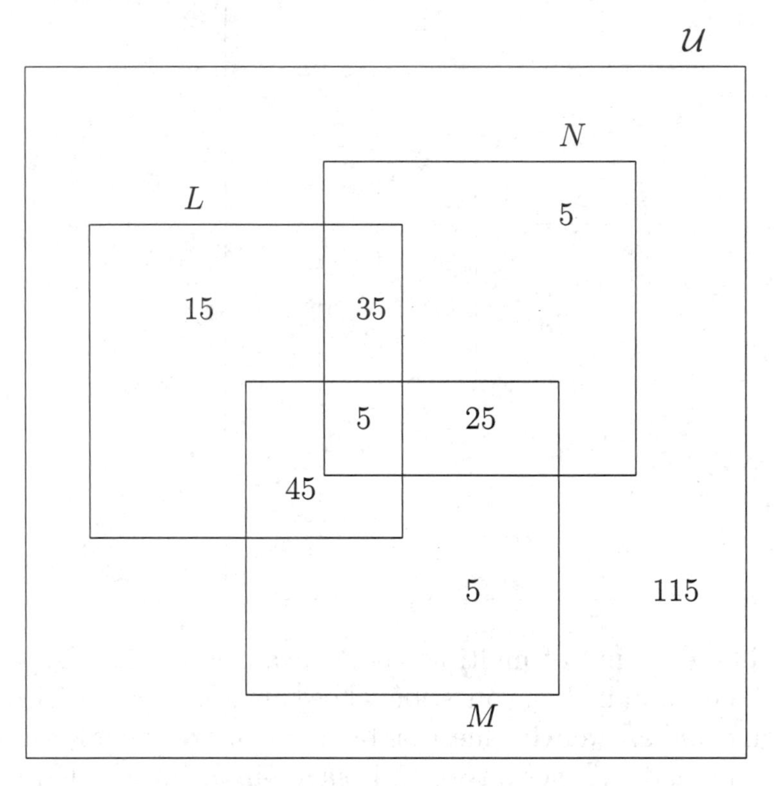

10. A survey of people reading polls during an election shows that 530 people believe that the Gallup polls accurately reflect the mood of the nation, 470 people believe that the Reuters polls accurately reflect the mood of the nation, 400 people believe that the Journal polls accurately reflect the mood of the nation. Two hundred and twenty five people believe that the Gallup polls and the Reuters Polls accurately reflect the mood of the nation, 175 people believe that the Gallup polls and the Journal polls accurately reflect the mood of the nation, and 120 people believe that the Reuters polls and the Journal polls accurately reflect the mood of the nation. One hundred people believe that all three polls reflect the mood of the nation. If each person surveyed answered, then how many people were surveyed.

11. Four hundred cat owners are surveyed. Two hundred own black cats, 170 own white cats, and 150 own bald cats. Also, 100 own black cats and white cats, 75 own black cats and bald cats, 50 own white cats and bald cats, and 25 own all three types of cat. Fill in the appropriate Venn diagram. Answer:

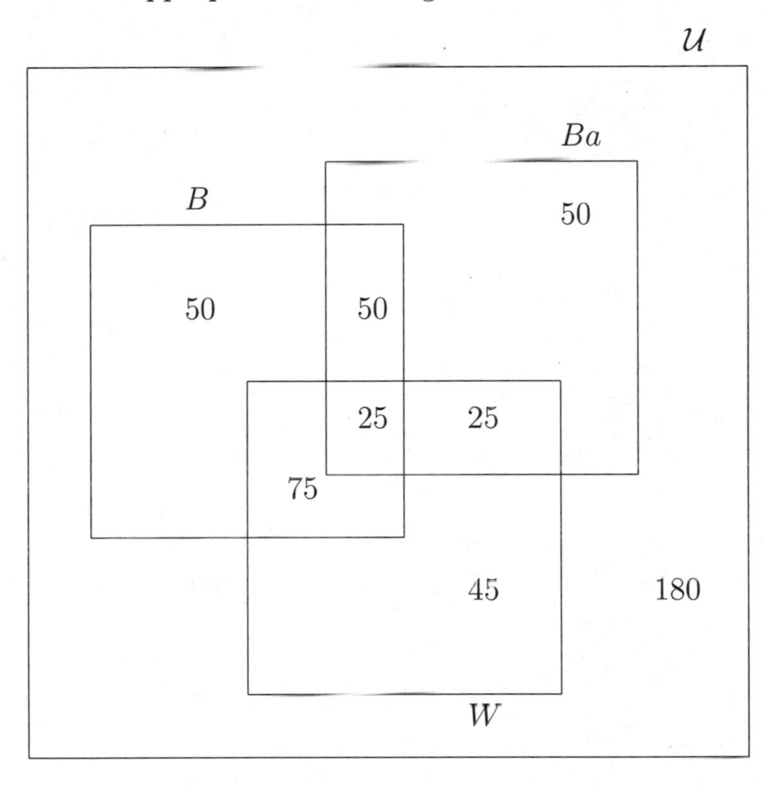

Chapter 4

Multiplication Principle

Some counting problems can be solved using a more systematic and more algebraic method than the structured arithmetic of the Venn diagram. These methods involve breaking a larger counting task into smaller more manageable counting problems. Over the next several chapters we will see that this method can be used in conjunction with more specialized counting techniques.

4.1 What Is the Principle?

EXAMPLE 4.1.1 Suppose that there is a series of tasks in front of us. To begin with we will assume that there are only two tasks in this series, task 1 and task 2. We wish to count the number of ways that the sequential task,

task 1 and then task 2

can be carried out. To further simplify our discussion, let us assume that task 1 can be done in three ways, and label the ways a, b, and c. Further suppose that task 2 can be done in two ways, labeled x and y. We will make a small chart diagramming the different ways that we can do task 1 and then task 2.

We can do *a first followed by* x. That sequence of tasks is denoted by ax. Working our way systematically through possible sequence of tasks, we would next execute the sequence of *a first and then y*. This sequence is listed as ay. We have written the first row of the following array:

$$\begin{array}{cc} ax & ay \\ bx & by \\ cx & cy \end{array}$$

The next sequence we would write is *b followed by x*. This is denoted by bx. The sequence *b and then y* is written as by. We have then written down the second row of the above array. In the last row of the array, cx represents the sequence *c and then x*, while cy denotes the sequence *c and then y*.

Getting back to our initial problem, we have shown that if task 1 can be done in three ways, and if task 2 can be done in three ways then the sequential task

task 1 and then task 2

can be done in $2 \cdot 3 = 6$ ways.

Let us abstract and expand the above example. Suppose you are given tasks S and T. Suppose that the first task S can be done in m ways, (where m is some whole number). Suppose that the second task T can be done in n ways (where n is some whole number). Then we can make lists of the ways in which S and T can be accomplished.

The outcomes of S are $a_1, a_2, \cdots, a_m$
The outcomes of T are $x_1, x_2, \cdots, x_n$

Then we make an array of all of the ways that we can do the sequential task *S and then T*.

$$m \left\{ \underbrace{\begin{array}{cccc} a_1x_1 & a_1x_2 & \cdots & a_1x_n \\ a_2x_1 & a_2x_2 & \cdots & a_2x_n \\ \vdots & \vdots & & \vdots \\ a_mx_1 & a_mx_2 & \cdots & a_mx_n \end{array}}_{n} \right.$$

We can see that the number of rows in the array is m by counting the subscripts $1, 2, \ldots, m$ of the a_i's. We can see that there are n

columns in the array by counting the subscripts of the x_j's. Thus, there are $m \cdot n$ entries or elements in the array. We have established the *Multiplication Principle.*

THE MULTIPLICATION PRINCIPLE 4.1.2 *Suppose that we are given tasks S and T, suppose that S can be done in m ways, and that T can be done in n ways. Then the sequential task*

S *and then* T

can be done in

$m \cdot n$

ways.

EXAMPLE 4.1.3 Suppose that you can travel from A to B in 6 ways, and from B to C in 4 ways. Then you can travel from A to B and then to C in $6 \cdot 4 = 24$ ways.

The multiplication principle also works when we are given more than just 2 tasks to accomplish.

EXAMPLE 4.1.4 Suppose that you have a bag of 13 different black chips, a bag of 13 different red chips, a bag of 13 different blue chips, and a bag of 13 different white chips. In how many ways can you choose a chip from bag 1, one from bag 2, one from bag 3, and then one from bag 4?

You are given 4 tasks. T_1 is to choose a black chip, T_2 is to choose a red chip, T_3 is to choose a blue chip, T_4 is to choose a white chip. Each bag has exactly 13 chips so in each task there are 13 different possible outcomes. Thus, the number of ways to accomplish the sequential task

$$T_1 \text{ and then } T_2 \text{ and then } T_3 \text{ and then } T_4$$

is

$$13 \cdot 13 \cdot 13 \cdot 13 = 13^4.$$

A slight variation on the previous problem yields the following.

EXAMPLE 4.1.5 Suppose you have three bags of chips. The first bag contains 26 diiferent white chips, the second bag contains 13 different red chips, and the third bag contains 10 different blue chips. In how many ways can you choose a chip from bag 1, a chip from bag 2, and a chip from bag 3?

You are given 3 tasks. T_1 is to choose a white chip, T_2 is to choose a red chip, and T_3 is to choose a blue chip. The number of ways to choose one white chip from the 26 given is 26, the number of ways to choose one red chip from 13 is 13, and the number of ways to choose one blue chip from 10 is 10. Thus, the number of ways to accomplish the sequential task

$$T_1 \text{ and then } T_2 \text{ and then } T_3$$

is

$$26 \cdot 13 \cdot 10.$$

In what follows *digits* are elements in the set $\{0, 1, 2, 3, 4, 5, 6, 7, 8, 9\}$. Multiple digit numbers like 999 and 10 are not digits.

EXAMPLE 4.1.6 A license plate in some state consists of a pair of digits followed by a single letter. To count the number of these plates, perform three tasks. In task T_1, choose a digit for the first place. In task T_2, choose a digit for the second place. For task T_3, choose a letter. The number of ways to choose a digit from $\{0, 1, 2, 3, 4, 5, 6, 7, 8, 9\}$ is 10, so the number of ways to do T_1 and T_2 is 10 each. The number of ways to choose a letter from $\{a, b, c, \ldots, z\}$ is 26 since there are 26 letters in the alphabet. Then by the multiplication principle the number of ways to do the sequential task

$$T_1 \text{ and then } T_2 \text{ and then } T_3$$

is

$$10 \cdot 10 \cdot 26.$$

EXAMPLE 4.1.7 A serial number consists of three letters followed by three digits. Because of printing considerations, the letters

I and O are not used and the numbers 1 and 0 are not used. How many serial numbers are there?

To count the serial numbers, perform two tasks several times. Let S be the task of choosing a letter other than I and O from the English alphabet. Then S can be done in 24 ways. Let T be the task of choosing a digit other than 1 and 0 from $\{0, 1, 2, 3, 4, 5, 6, 7, 8, 9\}$. Then T can be done in 8 ways.

Now, in task T_1, choose a letter other than I and O for the first place. In task T_2, choose a letter other than I and O for the second place. For task T_3, choose a letter other than I and O for the third place. Each of these tasks is S applied to a different place. Then each of the tasks T_1, T_2, T_3 can be done in 24 ways.

In task T_4, choose a digit other than 1 and 0 for the first digit place. In task T_5, choose a digit other than 1 and 0 for the second digit place. For task T_6, choose choose a digit other than 1 and 0 for the third digit place. Each of these tasks is T applied to a different place. Then each of the tasks T_4, T_5, T_6 can be done in 8 ways.

Thus, the process of making a serial number is the sequential task

$$T_1 \text{ and then } T_2 \text{ and then } T_3 \text{ and then } T_4 \text{ and then } T_5 \text{ and then } T_6.$$

To count the number of serial numbers, we simply count the number of ways to perform this sequential task. By the multiplication principle the number of ways to do this is

$$24 \cdot 24 \cdot 24 \cdot 8 \cdot 8 \cdot 8 = 8^3 \cdot 24^3.$$

There is one important problem that we will refer to often in the sequel.

EXAMPLE 4.1.8 Let $\{x_1, \ldots, x_t\}$ be an alphabet with t symbols. How many n-letter words are there in this alphabet?

There are t tasks to be performed. If $1 \leq i \leq n$ then Task i is to choose a letter for the i-the place in the word. In Task $i = T_i$ all we are doing is choosing a letter from the alphabet, which can be done in t ways since there are t different letters in the alphabet. Thus, the multiplication principle shows us that

$$\text{n}(T_1) \cdot \text{n}(T_2) \cdots \text{n}(T_n) = t^n.$$

Hence

The number of n-letter words from the alphabet with t letters $= t^n$

EXAMPLE 4.1.9 Let A be a set that contains exactly n elements. Let us use the multiplication principle to count the number of elements in the power set of A. That is, let us count the number of subsets of A.

Since A has exactly n elements we can write $A = \{1, \cdots, n\}$. Suppose that we are given a subset $X \subset A$. Then certain of the numbers i are in X and certain of the numbers i are not. We write a word W from the alphabet $\{0,1\}$. We write 1 in place 1 if $1 \in X$, and we write 0 in place 1 if $1 \notin X$. We write 1 in place 2 if $2 \in X$, and we write 0 in place 2 if $2 \notin X$. In general, we write 1 in place i if $i \in X$, and we write 0 in place i if $i \notin X$. This constructs our word W for X.

A example would look like this. Let $\mathcal{U} = \{1,2,3,4,5\}$, and let $X = \{2,4,5\}$. Then

$$X = \{2,4,5\} \qquad \begin{matrix} 1 & 2 & 3 & 4 & 5 \\ 0 & 1 & 0 & 1 & 1 \end{matrix} \qquad W = 01011$$

The top row of the table is the set A and the bottom row shows us how each 1 corresponds to a number in X. The 1's in the bottom row show us the numbers in X. Thus, $X = \{2,4,5\}$ corresponds to the word $W = 01011$.

Therefore, each subset $X \subset A$ corresponds to a word consisting of n 0's and 1's. Similarly, each sequence of 0's and 1's corresponds to a subset X of A. So to count the subsets of A we can instead count the number of words consisting of n bits from $\{0,1\}$.

To count the number of words W containing n bits from $\{0,1\}$ we have n tasks to complete. Each task is *fill in a place.* There are n places to fill in W, and each can be filled in 2 ways: with a 0 or a 1. The multiplication principle shows us that this sequence of n tasks can be done in $\underbrace{2 \cdots 2}_{n} = 2^n$ ways. Thus, the total number of words containing exactly n symbols from $\{0, 1\}$ is 2^n. But these words correspond to the subsets of A, so we have counted

$$\text{the number of subsets of } \{1, 2, \cdots, n\} = 2^n.$$

4.2 Exercises

1. Tina has a closet containing 10 pairs of shoes, 8 scarves, 15 dresses, and today she will have a choice of 4 different hairstyles. In how many ways can she choose an outfit and makeup her hair for her business day? Answer: $10 \cdot 8 \cdot 15 \cdot 4$

2. Hillary will make business stops in three major cities in one day. She can only talk with 1 of 25 people in city 1, she can only talk with 1 of 10 board members in city 2, and she can only talk with 1 of 17 women in city 3. In how many different ways can she fill her speaking engagements that day? Answer: $25 \cdot 10 \cdot 17$

3. In how many ways can you make a row consisting of the 4 different chips labeled A, B, C, D? Answer: $4 \cdot 3 \cdot 2 \cdot 1$

4. You are holding a book with 200 pages, and you randomly open the book to a page. In how many ways can you open the book 5 times? Answer: 200^5

5. You are holding a book with 200 pages, and you randomly open the book to a page. In how many different ways can you open the book 5 times? Answer: $200 \cdot 199 \cdot 198 \cdot 197 \cdot 196$

6. Choose a card from a standard deck of 52 cards, and replace it in the deck. Do this 3 times. In how many ways can you do this? Answer: 52^3

7. Choose a card from a standard deck of 52 cards, and remove it from the deck. Do this 3 times. In how many ways can you do this? Answer: $52 \cdot 51 \cdot 50$

Chapter 5

Permutations

A *word* is a string of symbols from an alphabet. These words do not have to be found in a dictionary. They are just strings of letters from some alphabet.

EXAMPLE 5.0.1 1. The strings *abc* and *pxqq* are words in the English language.

2. The strings 00000, 0123, and 191 are words in the alphabet of digits $\{0, 1, 2, 3, 4, 5, 6, 7, 8, 9\}$.

3. The strings 0101010 and 111 are words in the binary alphabet $\{0, 1\}$.

The counting method we now introduce works especially well when counting words in an alphabet.

A *permutation* is a word from some alphabet with the property that no two symbols in the word are alike. We say that there are *no repeated symbols.* In a permutation, the order in which the symbols appear is significant because these are words. After all, *hits* and *this* are different words even though they have the same letters. The words 9995 and 5999 are clearly different if you think of them as your salary.

EXAMPLE 5.0.2 1. The words *cba*, *bca*, and *abc* are permutations since no two symbols in *abc* are alike. We would say that *cba* and *bca* are *arrangements of abc*.

2. 01 and 9745 are permutations from the alphabet of digits.

3. *aaaa* and 887123 are not permutations in any alphabet.

In this chapter we will learn how to count permutations.

5.1 Some Special Numbers

To count the number of permutations that there are, we will use some important numbers. The following numbers are called *factorial.*

$0! = 1, 1! = 1,$ and for each whole number $n \geq 2$, $n! = n(n-1)!$.

For instance, $2! = 2 \cdot 1! = 2 \cdot 1 = 2$, and $3! = 3 \cdot 2! = 3 \cdot 2 = 6$. In general, we will write

$$n! = n \cdot (n-1) \cdot (n-2) \cdots 2 \cdot 1$$

when we want to calculate n factorial. Thus, $5! = 5 \cdot 4 \cdot 3 \cdot 2 \cdot 1$ and $10! = 10 \cdot 9 \cdot 8 \cdot 7 \cdot 6 \cdot 5 \cdot 4 \cdot 3 \cdot 2 \cdot 1$. We do not really care what these numbers are on a calculator, just that 5! is a product of positive whole numbers less than 5.

The numbers used to count permutations are so important that we have a special notation for them. Let $n \geq k > 0$ be whole numbers. Then

$$P(n, k) = \frac{n!}{(n-k)!}.$$

Notice that $P(n, k)$ has no meaning as yet. It is just a simple way of expressing a certain number.

EXAMPLE 5.1.1 1. $P(3,2) = \frac{3 \cdot 2 \cdot 1}{(3-2)!} = \frac{3 \cdot 2 \cdot 1}{1!} = 3 \cdot 2 = 6.$

2. $P(5,3) = \frac{5 \cdot 4 \cdot 3 \cdot 2 \cdot 1}{(5-3)!} = \frac{5 \cdot 4 \cdot 3 \cdot 2 \cdot 1}{2 \cdot 1} = 5 \cdot 4 \cdot 3 = 60.$

3. $P(100,40) = \frac{100!}{(100-40)!} = \frac{100!}{60!}$, and that is as far as you should be required to carry out that calculation.

EXAMPLE 5.1.2 1. $P(n,n) = \frac{n!}{(n-n)!} = \frac{n!}{0!} = \frac{n!}{1} = n!.$

2. $P(n,0) = \frac{n!}{(n-0)!} = \frac{n!}{n!} = 1.$

EXAMPLE 5.1.3 1. $P(2,3)$ is *undefined* since $2 < 3$.

2. $P(\frac{1}{2},2)$ is *undefined* since $\frac{1}{2}$ is not a whole number.

3. $P(0,0) = \frac{0!}{0!} = 1.$

Observe that $P(n,k)$ is non-zero in our calculations. Now that we have the right numbers we can count permutations.

5.2 Permutations Problems

A *permutation problem* is a problem in which you are asked to count the number of permutations having k symbols in some alphabet of n symbols. The solution to a permutation problem is $P(n,k)$. Note the use of n and k. Let us show why that is True.

EXAMPLE 5.2.1 The permutations we are counting have k places or symbols in them. We construct them in a sequence of tasks. T_1 = choose the first symbol from the alphabet of n. T_2 = choose the second symbol from the alphabet of $n-1$ symbols. (Since we have already used a letter and since we cannot repeat symbols, there are $n-1$ symbols to choose from.) T_3 = choose the third symbol from the alphabet of $n-2$ symbols. (Since we have already used two letters and since we cannot repeat symbols, there are $n-2$ symbols to choose from.) Continue inductively. T_k = choose the k-th symbol from the alphabet of $n-k+1$ symbols. (Since we have already used $k-1$ letters and since we cannot repeat symbols, there are $n-(k-1) = n-k+1$ symbols to choose from.) Then by the multiplication principle the sequential task of making a permutation with k symbols from an alphabet of n symbols is

$$n(n-1)(n-2)\cdots(n-k+1).$$

Let us compare this number to $P(n,k)$. **As long as $k \neq 0$ the formula for $P(n,k)$ will yield**

$$\begin{aligned} P(n,k) &= \frac{n!}{(n-k)!} \\ &= \frac{n\cdot(n-1)\cdots(n-k+1)\cdot[(n-k)\cdot(n-k-1)\cdots 2\cdot 1]}{[(n-k)\cdot(n-k-1)\cdots 2\cdot 1]} \\ &= n\cdot(n-1)\cdots(n-k+1) \end{aligned}$$

The numbers are the same. Thus,

given integers $n \geq k > 0$, the number of permutations of k symbols from an alphabet of n symbols is

$$P(n,k) = n\cdot(n-1)\cdots(n-k+1).$$

Notice that there are k factors in the product $P(n,k)$. But also note that

$P(n,0) = 1$ the number of permutations of $\emptyset$ is 1.

Please keep this observation in mind during your reading.

EXAMPLE 5.2.2 How many permutations of 10 letters are there from the alphabet $\{a,b,c,d,e,f,g,h,i,j\}$?

To answer this, note that the alphabet has 10 letters in it. Thus, this is a problem in which you are asked to count the number of permutations having 10 letters in an alphabet of 10 letters. This is exactly a permutation problem. There are $P(10, 10) = 10!$ such permutations.

EXAMPLE 5.2.3 How many permutations of 3 letters are there from the English alphabet?

The English alphabet has 26 letters in it. You are asked to count the number of permutations having exactly 3 letters in an alphabet of 26 symbols. This is exactly a permutation problem. There are $P(26, 3) = 26 \cdot 25 \cdot 24$ such permutations.

In the event that it is not clear that we are dealing with a permutation, we can employ the following equivalent criteria. A *permutation problem* is characterized as follows:

1. There is one alphabet of n symbols from which the objects to be counted are made.

2. No repeated symbols are allowed in the k symbol permutations to be counted.

3. The order in which the symbols appear is significant.

Given this information about the problem then we are solving a permutation problem. The number of permutations is then $P(n, k)$.

EXAMPLE 5.2.4 How many words of 5 different letters are there from the alphabet $\{a, e, i, o, u, y\}$?

There is one alphabet of 6 letters. The problem dictates that the 5 letters used are different. Since we are counting words, order is significant. The three criteria are thus met. This is a permutation problem. Hence, there are $P(6, 5) = 6 \cdot 5 \cdot 4 \cdot 3 \cdot 2 = 6!$ such words.

EXAMPLE 5.2.5 How many words of 3 different digits are there from the alphabet $\{0, 2, 4, 6, 8\}$?

There is one alphabet of 5 digits. The problem dictates that the 3 digits used are different. Since we are counting words, order is significant. The three criteria are thus met. This is a permutation problem, and there are $P(5, 3) = 5 \cdot 4 \cdot 3$ such words.

EXAMPLE 5.2.6 How many words of 3 different bits are there from the alphabet $\{0, 1\}$?

There is one alphabet of 2 letters. The problem dictates that the 3 bits used are different. But since there are only 2 letters, a 3 letter word must contain a repeated symbol. Hence this is not a permutation problem.

If we remove the requirement that the letters have to be different, this problem can be solved using the multiplication principle. There are 3 places, and there are exactly 2 letters to be used in each place. Then $T_1 = T_2 = T_3 =$ *choose a letter for the place.* Each of these tasks T_1, T_2, T_3 can be done in 2 ways. By the multiplication principle there are $2 \cdot 2 \cdot 2 = 2^3$ words with 3 letters from the alphabet $\{0, 1\}$.

EXAMPLE 5.2.7 License plates in some state consist of 3 different digits followed by 3 different letters. How many of these plates are there?

It seems that the problem states that the digits used are different, and they are. It seems that we are counting the number of words with 6 symbols, and we are. But to fill out the word or license plate, we need to dip into 2 alphabets. One alphabet is the set of digits and the other is the set of English letters. Hence this is not a permutation problem.

It can be solved using a multiplication problem, though.

The first problem is to count the number of words with 3 different symbols from the alphabet of 10 digits. This is exactly what we called a permutation problem above. The number of such words is then $P(10, 3) = 10 \cdot 9 \cdot 8$.

The second problem is to count the number of words with 3 different symbols from the alphabet of 26 English letters. This is exactly what we called a permutation problem above. The number of such words is then $P(26, 3) = 26 \cdot 25 \cdot 24$.

Hence, the multiplication principle shows us that there are

$$P(10, 3)P(26, 3)$$

license plates constructed in the stated way.

5.3 Exercises

1. Find $P(4,4)$, $P(4,2)$, $P(n,2)$, and $P(n+1,n)$. Answer: $P(4,4) = 4!$, $P(4,2) = 4{\cdot}3$, $P(n,2) = n(n-1)$, $P(n+1,n) = (n+1)!$

2. How many permutations having 5 different digits from the alphabet $\{0,1,2,3,4,5,6,7,8,9\}$ of digits are there? Answer: $P(10,5)$

3. How many words of 2 different symbols are there from an alphabet having n symbols? Answer: $P(n,2) = n(n-1)$

4. How many words of 30 different symbols are there from an alphabet having 100 symbols? Answer: $P(100,30)$.

5. How many words of k different symbols are there from an alphabet having n symbols? Answer: $P(n,k)$.

6. How many arrangements of the word *abcde* are there? Answer: 5!

7. How many arrangements of the word 012345 are there? Answer: 6!

8. How many license plates have 5 different digits other than 0 and 1 followed by 3 English letters other than I and O? Answer: $P(8,5)P(24,3)$

9. How many two-word sentences in English consist of a 4 letter word and a 3 letter word? Answer: 26^7

10. How many two word sentences in English consist of a word with 4 different letters followed by a word of 3 different letters? Answer: $P(26,4)P(26,3)$

11. A professor has a class of 35 students in 50 chairs. To emphasize the fact that he does not take attendance, he continually seats the students so that some of them are sitting in different seats each day. Count the number of ways he can seat the class. There are 3 meetings a week and 12 weeks in the semester. Can the professor change the seating throughout the semester without repeating a seating? Answer: $P(50,35)$. This number is much larger than the the 36 meetings the professor has a semester.

Chapter 6

Combinations

In listing the elements in a set, we have ignored an important property. The order in which we list the elements of a set is not significant. In other words,

$$\{1,2,3,4\} = \{1,3,2,4\} = \{4,3,2,1\}$$
$$\{a,b,9,10\} = \{b,a,9,10\} = \{a,10,b,9\}.$$

Furthermore, we do not repeat an element in a set when listing it. So $\{1\}$ is written instead of $\{1,1\}$. The set $\{M,I,S,S,I,P,P,I\}$ is not written. Instead we write $\{M,I,S,P\}$. Let me repeat. We do not repeat elements when writing a set as a list of elements. This will be important when we start to count elements in sets.

6.1 Some Special Numbers

It will be useful to know shorthand notation for some of the numbers that we will encounter when we count combinations of elements. Recall that $n! = n \cdot (n-1) \cdots 2 \cdot 1$ for whole numbers $n \geq 2$ and that $1! = 0! = 1$.

Let $n \geq k \geq 0$ be whole numbers. The combination number n *choose* k is

$$\binom{n}{k} = \frac{n!}{k!(n-k)!} = \frac{n \cdot (n-1) \cdots (n-k+1)}{k!}.$$

EXAMPLE 6.1.1 1. $\binom{3}{2} = \frac{3 \cdot 2}{2 \cdot 1} = 3$

2. $\binom{5}{3} = \frac{5 \cdot 4 \cdot 3}{3 \cdot 2 \cdot 1} = 10$

EXAMPLE 6.1.2 1. Recall that by definition $0! = 1$. Then $\binom{n}{0} = \frac{n!}{0!(n-0)!} = \frac{n!}{1 \cdot n!} = 1$

2. Recall that $n! = n \cdot (n-1)!$ for whole numbers $n \geq 1$. Then $\binom{n}{1} = \frac{n!}{1!(n-1)!} = \frac{n \cdot (n-1)!}{(n-1)!} = n$

EXAMPLE 6.1.3 1. $\binom{2}{3}$ is undefined. There is no numerical value equal to this symbol. The formula given for n choose k is only good for numbers such that $n \geq k$.

2. $\binom{2}{1/2}$ is undefined.

Now that we have these special numbers we can examine what a combination problem is.

6.2 Combination Problems

The prototypical *combination problem* is the following. Let $n \geq k \geq 0$ be whole numbers. *In how many ways can we choose k elements from a set of n elements?* In other words, *in how many ways can we choose k of n elements in a set?*

> The number of different ways to
> choose k elements from a set of n different elements is
> $$\binom{n}{k}.$$

In the next example we will prove that this is indeed the case.

EXAMPLE 6.2.1 We are asked to choose k elements from a set of n elements. Let us use $C(n, k)$ as a symbolic shorthand that denotes the number of ways to make this choice. Thus $C(n, k)$ is the number of ways to choose k elements from n elements.

To count we begin with a problem that we know. The number of permutations of k symbols from an alphabet of n symbols is $P(n, k)$. Another way to count this number is as follows. The first task is to choose k symbols from the set of n symbols that is the alphabet. By our definition of $C(n, k)$, this can be done in $C(n, k)$ ways. Since we chose these symbols from the alphabet, we have k *different* symbols. From our work in the previous chapter, the number of permutations of k different symbols is $k!$. Then by the multiplication principle, the number of ways to choose k elements from a set of n elements is

$$P(n, k) = C(n, k)k!.$$

Substituting in the value for $P(n, k)$ from page 66 in the previous chapter we have

$$\frac{n!}{(n-k)!} = C(n, k)k!$$

and dividing by $k!$ yields

$$\frac{n!}{k!(n-k)!} = C(n, k).$$

But the left-hand side of this equation is $\binom{n}{k}$ so

$$\binom{n}{k} = C(n, k).$$

Therefore, $\binom{n}{k}$ is the number of ways to choose k elements from n elements in a set. This is what we set out to do.

Let us exploit this formula to count some solutions to combination problems.

EXAMPLE 6.2.2 In how many ways can we choose 10 elements from a set of 12 elements? This is exactly the problem we began this chapter with. The answer is $\binom{12}{10} = \frac{12 \cdot 11}{2} = 66$.

EXAMPLE 6.2.3 In how many ways can we choose 6 letters from the English alphabet? This problem is the same as choosing 6 elements from a set of 26 elements, which can be done in $\binom{26}{6} = \frac{26!}{6!20!}$ ways. We have no interest in a further reduction of this fraction.

EXAMPLE 6.2.4 We have a bag of 100 different chips. In how many ways can we take a handful of 40 chips from the bag? This problem is the same as choosing 40 elements from a set of 100 elements, which can be done in $\binom{100}{40} = \frac{100!}{40!60!}$ ways.

EXAMPLE 6.2.5 We have a shelf of 25 different books. In how many ways can we remove 13 books from the shelf? This problem is the same as choosing 13 of 25 elements in a set, which can be done in $\binom{25}{13} = \frac{25!}{13!12!}$ ways.

Now let us solve some more complex problems. When I refer to a *standard deck of cards* I will mean a deck of 52 cards that can be assorted as follows. There are four suits in the deck: hearts, diamonds, clubs, spades. Thirteen of the cards are hearts, and they are labeled $2, 3, 4, 5, 6, 7, 8, 9, 10, J, Q, K, A$. Thirteen of the cards are diamonds, and they are labeled $2, 3, 4, 5, 6, 7, 8, 9, 10, J, Q, K, A$. Thirteen of the cards are clubs, and they are labeled $2, 3, 4, 5, 6, 7, 8, 9, 10, J, Q, K, A$. Thirteen of the cards are spades, and they are labeled $2, 3, 4, 5, 6, 7, 8, 9, 10, J, Q, K, A$. A *kind* of card is from the four cards with one of the labels $2, 3, 4, 5, 6, 7, 8, 9, 10, J, Q, K, A$. When we say that we have *two of a kind* we are saying that we have two cards labeled in the same way. They will be of different suits. While there are four aces (one from each of the suits), the deck does not repeat cards. A *hand* is a choice of cards from the deck. Therefore, a deck of cards is a set of 52 elements, numbered and labeled in such a way that the 52 cards are different.

We will not engage in games. These are counting problems. Thus you will find us doing what looks like cheating when we solve these problems.

EXAMPLE 6.2.6 In how many ways can we choose a card from a deck? This is the same as choosing one element from a set of 52 elements. This can be done in $\binom{52}{1} = 52$ ways.

EXAMPLE 6.2.7 In how many ways can we choose an Ace from a deck? In this problem we are not choosing an Ace from the deck, we are choosing an Ace from the four Aces that exists in the deck. This is the same as choosing one element from a set of 52 elements. This can be done in $\binom{4}{1} = 4$ ways. Let us examine that answer. Suppose you take a deck and begin choosing Aces from the deck. You cannot choose more than the four Aces that are in the deck. Specifically, you do not have $\binom{52}{1} = 52$ ways to choose that Ace! You have 4.

EXAMPLE 6.2.8 When choosing a specific card from a deck, you are choosing that card from the 4 others of the same kind. Let us choose one card of the kind C from a deck. The kind C represents one of the 13 kinds of cards in a deck. There are exactly 4 cards of the kind C in the deck. Thus we are choosing one card of the kind C from 4 cards of that kind. We choose 1 of 4 elements in a set. This can be done in $\binom{4}{1} = 4$ ways. Thus the number of ways of choosing one card of kind C from a standard deck is always 4.

EXAMPLE 6.2.9 In how many ways can we choose 5 cards from a deck? This is the same as choosing 5 elements from a set of 52 elements. This can be done in $\binom{52}{5}$ ways. This is a lot of hands with 5 cards without repeating hands.

EXAMPLE 6.2.10 We are given a bag of 20 different colored chips. There are 12 red chips and 13 blue chips.

1. In how many ways can we choose 6 red chips from the bag? This is the same as choosing 6 elements from a set of 12 elements. This can be done in $\binom{12}{6}$ ways. Since we are not choosing blue chips we ignore them and choose only from the red chips. That is why there are $\binom{12}{6}$ ways to choose 6 red chips from the bag.
2. In how many ways can we choose 7 blue chips from the bag? In $\binom{13}{7}$ ways.

Notice that in this problem it is perfectly alright to ignore the blue chips when we choose the red chips. And yet, when choosing an Ace from a deck, some of you are still not sure that we should ignore the other cards.

EXAMPLE 6.2.11 We are given a bag of 35 different colored chips. There are 20 black chips and 15 white chips. In how many ways can we choose 5 black chips and 6 white chips from the bag? The solution requires the multiplication principle.

The first task is to choose 5 of 20 black chips in the bag. This is the same as choosing 5 elements from a set of 20 elements, which can be done in $\binom{20}{5}$ ways.

Task two is to choose 6 of 15 white chips in the bag. This is the same as choosing 6 elements from a set of 15 elements, which can be done in $\binom{15}{6}$ ways.

By the multiplication principle, the total number of ways to make your choice is

$$\binom{20}{5}\binom{15}{6}.$$

No further simplification is needed.

6.3 Exercises

1. In how many ways can we choose 5 elements from a set of 7 elements? Answer: $\binom{7}{5}$

2. In how many ways can we choose 2 of n elements in a set? Answer: $\binom{n}{2}$

3. Argue why $\binom{n}{0} = 1$.

4. In how many ways can we choose 3 letters from the English alphabet? Answer: $\binom{26}{3}$

5. In how many ways can we choose 3 cards from a deck? Answer: $\binom{52}{3}$

6. We are given a bag of 26 different colored chips. There are 13 red chips and 13 black chips.

   (a) In how many ways can we choose 5 red chips from the bag ? Answer: $\binom{13}{5}$

   (b) In how many ways can we choose 3 black chips from the bag? Answer: $\binom{13}{3}$

7. In how many ways can we choose two Aces from a deck? Answer: $\binom{4}{2}$

8. In how many ways can you choose an Ace, a King, a Queen, a Jack, and a 10 from a deck? Answer: $\binom{4}{1}^5 = 4^5$

Chapter 7

Problems Combining Techniques

The purpose of this chapter is to solve counting problems that are neither permutation nor combination problems. They are a blending of the multiplication problem and some other techniques. To solve these problems we have to decompose them into smaller more manageable problems. These smaller problems will resemble the permutation and combination problems that we solved in the previous chapters.

7.1 Significant Order

License plates are a good source of problems that are not permutation problems, but for which the order of the symbols is significant.

EXAMPLE 7.1.1 In how many ways can we form a license plate that consists of 4 digits followed by 3 letters?

In problems like this, deconstruct the object considered, in this case a license plate, into a sequence of simpler problems or objects.

For this license plate, let T_1 be the task of constructing a sequence of 4 digits. Then T_1 is accomplished by choosing a digit for each of four places, and each of these tasks is done in 10 ways. (There are 10 digits.) Then the ways to do T_1 is 10^4.

Next, let T_2 be the task of constructing a sequence of 3 letters. Then T_2 is accomplished by choosing a letter for each of the 3 places, which in each place can be done in 26 ways, so T_2 can be done in 26^3 ways.

Thus, the license plate, which is the outcome of doing the sequential task T_1 and then T_2, is constructed in $10^4 \cdot 26^3$ ways.

Note the difference between the above problem and the next problem. This difference is that in the next problem we have *different* digits and letters. This changes the formulas used to solve the problem significantly.

EXAMPLE 7.1.2 In how many ways can we form a license plate that consists of 4 different digits followed by 3 different letters?

The problem decomposes into tasks T_1 and T_2. In T_1 construct a sequence (a word) of 4 different digits. This is the prototypical permutation problem, so it can be done in $P(10, 4)$ ways. In T_2 construct a sequence (a word) of 3 different letters. This is also a prototypical permutation problem, which is done in $P(26, 3)$ ways. The license plate is constructed by doing the sequential task T_1 and then T_2, which can be done in $P(10, 4)P(26, 3)$ ways.

EXAMPLE 7.1.3 In how many ways can we form a license plate that consists of 4 different digits not equal to 0 or 1, followed by 3 different letters not equal to I or O?

There are two tasks. In T_1 delete 0 and 1 from the alphabet to be used, and construct a sequence (a word) of 4 different digits. This is the prototypical permutation problem from an alphabet of 8 symbols. This can be done in $P(8, 4)$ ways. In T_2 delete I and O from the alphabet of 26 letters and construct a sequence (a word) of 3 different letters. This is also a prototypical permutation problem from an alphabet of 24 letters. This can be done in $P(24, 3)$ ways. The license plate is constructed by doing the sequential task T_1 and then T_2. This is done in $P(8, 4)P(24, 3)$ ways.

7.2 Order Not Significant

In the following problems utilize the following setting. There is a bag of different colored chips. Thirteen of the chips are red chips numbered 1 through 13. Thirteen of the chips are black chips numbered 1 through 13. Thirteen of the chips are blue chips numbered 1 through 13. Thirteen of the chips are white chips numbered 1 through 13. There are a total of 52 chips in the bag.

EXAMPLE 7.2.1 In how many ways can we choose 13 chips from the bag?

There are 52 different chips in the bag and we choose 13 of them. This is the prototypical combination problem, which is done in $\binom{52}{13}$ ways.

When we choose a handful of chips from the bag we can cheat and think of choosing first from one color and then another. In this way we construct the handful one color at a time. For instance, the number of ways to choose 5 red chips from the bag is found by choosing the 5 chips from the red chips. This can be done in $\binom{13}{6}$ ways.

EXAMPLE 7.2.2 In how many ways can we choose a handful of 5 red chips and 6 black chips from the bag?

Break the problem into two smaller tasks. Choose 5 of 13 different red chips from the bag. This is the problem of choosing 5 elements from a set of 13 elements, which is done in $\binom{13}{5}$ ways.

Choose 6 of 13 different black chips from the bag. This is the problem of choosing 6 elements from a set of 13 elements, which is done in $\binom{13}{6}$ ways. Then the number of ways to choose the handful of 5 red and 6 black chips from the bag is

$$\binom{13}{5}\binom{13}{6}$$

ways.

Instead of using the colors to make a counting problem, let us use the numbers on the chips to make a different counting problem.

From our description of the bag, there is a chip numbered n for each color. Thus, there are four chips numbered 13, and the same applies to each of the numbers $1, 2, 3, 4, 5, 6, 7, 8, 9, 10, 11, 12, 13$.

When you are choosing chips with the same number on them, say the number 2, then you choose from the set of chips with number 2. By the statement of the problem, you are choosing chips numbered 2. You ignore the other chips in the bag and choose a 2 from the set {red 2, black 2, blue 2, white 2}.

EXAMPLE 7.2.3 In how many ways can you choose 1 chip numbered 3? In how many ways can you choose 2 chips numbered 3?

We are choosing 1 chip from a set of 4 chips numbered 3. This is a combination problem and it is done in $\binom{4}{1} = 4$ ways. Actually, the numbers here are so small that we can show all of the ways to choose a 3 from the bag. Your only possible choices are to choose a red 3, a black 3, a blue 3, and a white 3. That is 4 ways in anyone's book.

In the second question we are choosing 2 chip from a set of 4 chips numbered 3. This is done in $\binom{4}{2}$ ways.

Here is an interesting and useful idea. When we choose $n - 1$ elements from a set of n elements, we can do this in $\binom{n}{n-1}$ ways. But we can also choose the set of $n - 1$ elements by choosing instead the 1 element that we will leave behind. This leaving is done in $\binom{n}{1} = n$ ways. Therefore

$$\binom{n}{n-1} = \binom{n}{1} = n$$

for whole numbers $n > 0$.

EXAMPLE 7.2.4 In how many ways can you choose 3 chips numbered 13 and 2 chips numbered 12 from the bag in one handful?

We will fill out this handful by doing 2 tasks. First choose 3 chips numbered 13. This is the same as choosing 3 elements from a set of 4 elements, which is done in $\binom{4}{3} = 4$ ways.

Next choose 2 chips numbered 12. This is the same as choosing 2 elements from a set of 4 elements, which is done in $\binom{4}{2}$ ways.

The total number of handfuls is then fund by doing the sequential task T_1 and then T_2, which is done in

$$\binom{4}{3}\binom{4}{2} = 4 \cdot \frac{4 \cdot 3}{2} = 24$$

ways.

EXAMPLE 7.2.5 In how many ways can you choose 3 chips with one number and 2 chips with a different number?

This problem requires that we choose the numbers on the chips and then worry about which chips we choose. First choose the numbers to appear on the 3 and the 2 chips chosen. This is the same as choosing a permutation of 2 symbols from the alphabet of $\{1, 2, 3, 4, 5, 6, 7, 8, 9, 10, 11, 12, 13\}$, which can be done in $P(13, 2)$ ways. (The order matters here since one number corresponds to 3 chips and the other corresponds to 2 chips.) Say we have chosen a and b.

Choose 3 of the 4 chips numbered a. This is done in $\binom{4}{3}$ ways.

Next choose 2 of the 4 chips numbered b. This is done in $\binom{4}{2}$ ways.

Thus, the number of ways to choose 3 chips with one number and 2 chips of another number is

$$P(13, 2)\binom{4}{3}\binom{4}{2} = 4^2 \cdot 13 \cdot 12 \cdot 3$$

ways.

EXAMPLE 7.2.6 In how many ways can you choose 2 chips with one number and 2 chips with a different number, and one chip of an other number?

Our experience shows that no matter what numbers have been chosen, the number of ways to choose 2 of 4 of a numbered chip is $\binom{4}{2}$ ways. We do this twice, once for each number on the chips.

In counting the numbered chips in this problem, we observe that there is no way to distinguish the difference between a pair of chips numbered C and a pair of chips numbered D. You might suggest that we choose one number first and then the second, but that implies that you are choosing your chips one at a time. In these problems, you choose the chips at once. In this way the chips come to you at the same time. The order that they come to you is not defined. They came to you at the same moment. Thus, the order in which we choose the two numbers is not significant. In choosing the 2 numbers to be used in this problem, we are simply choosing 2 numbers from the alphabet $\{1, 2, 3, 4, 5, 6, 7, 8, 9, 10, 11, 12, 13\}$. This is done in $\binom{13}{2}$ ways.

Next, choose a different numbered chip from the $52 - 8 = 44$ chips left in the bag. We ignore the 8 numbered chips that we chose because the problem asks for a different numbered chip. There are 44 ways to do this.

Therefore, the choice of 2 of one numbered chip, 2 of another numbered chip, and 1 of a different numbered chip can be done in

$$\binom{4}{2}\binom{4}{2}\binom{13}{2}44$$

ways.

It is traditional to count poker hands when the course reaches this stage of development. To do this we will need to know what a standard deck of cards is. We gave such a description on page 76. A poker hand is a subset of the standard deck consisting of 5 cards. The cards are from the deck, so the cards are different in the same way that the numbered chips in our bag are different. The chips have different colors and different numbers. The cards are different because they have different *suits* and different symbols on them. In fact, each of the problems we did with the bag of chips is the same as a problem about cards. We will use this fact in the following example.

EXAMPLE 7.2.7 The word *deck* in this problem refers to the standard deck of 52 different cards.

1. The number of ways to choose a poker hand from the deck is the same as the number of ways to choose 5 chips from the bag. We are choosing 5 cards from a set of 52. This is done in
$$\binom{52}{5}$$
ways.

2. The number of ways to choose 3 hearts and 2 diamonds from the deck is the same as the number of ways to choose 3 from 13 elements in a set and 2 of 13 elements in a set. These are done in $\binom{13}{3}$ and $\binom{13}{2}$ ways, respectively. Then the 3 and 2 can be chosen in exactly
$$\binom{13}{3}\binom{13}{2}$$
ways.

EXAMPLE 7.2.8 The word *deck* in this problem refers to the standard deck of 52 different cards.

1. The number of ways to choose 3 Aces and 2 Kings from the deck is the same as the number of ways of choosing 3 chips numbered 13 and 2 chips numbered 12. This was counted in Example 7.2.4. The answer is
$$\binom{4}{3}\binom{4}{2}$$
ways.

2. Count the number of ways to choose 3 of one kind, (a kind refers to the symbol on the card), and 2 of a different kind. This problem is the same as choosing 3 chips from our bag with the same number on them, and then choosing 2 chips with the same number on them. This problem is done in Example 7.2.5, where the answer is given as

$$13 \cdot 12 \cdot \binom{4}{3}\binom{4}{2}$$

ways.

3. Count the number of five cards hands that have 2 of one kind, 2 of another kind, and one of a different kind. This problem is the same as choosing a handful of chips from our bag such that 2 have one number on them, 2 have a different number on them, and a third chip with a different number on it. This problem is solved in Example 13.2.5, where the answer given is

$$44\binom{13}{2}\binom{4}{2}\binom{4}{2}.$$

Note again that Example 7.2.8(2) and 7.2.8(3) count differently. In Example 7.2.8(2), when we make the choice of kinds of cards to label the 3 of a kind and the 2 of a kind, we choose two symbols, say a, b. The reason that their order is significant is this. Having 3 of the kind a and 2 of the kind b is different from having 3 of the kind b and 2 of the kind a. Since a, b and b, a yield different hands, the order is significant in this part of the problem. A permutation number is then called for.

In Example 7.2.8(3), when we make the choice of kinds of cards to label the 2 of a kind, we choose two symbols, say a, b, from 13 symbols. The reason that their order is *not* significant is this. By choosing 2 of one kind and 2 of another kind we are not choosing in an order. The pairs of cards just exist in our hands. There is no way to differentiate them. Thus, order is not significant in this part of the problem. A combination number is then called for.

7.3 Exercises

The *bag* refers to our bag given on page 139. A *deck* of cards refers to a standard deck, as given on page 76.

1. Count the number of handfuls from the bag consisting of 4 of one numbered chip and 3 of another numbered chip. Answer: $13 \cdot 12 \cdot 4$.

2. Count the number of handfuls from the bag consisting of 3 of one numbered chip and 3 of another numbered chip. Answer: $4^2 \cdot \binom{13}{2}$

3. Count the number of handfuls from the bag consisting of 4 of one numbered chip, 3 of another numbered chip, and 2 of another number chip. Answer: $13 \cdot 12 \cdot 11 \cdot 4 \cdot \binom{4}{2}$

4. Count the number of hands from the deck consisting of 4 of one kind and 3 of another. Answer: $13 \cdot 12 \cdot 4$

5. Count the number of hands from the deck consisting of 3 of one kind and 3 of another kind. Answer: $4^2 \cdot \binom{13}{2}$

6. Count the number of hands from the deck that consist of 2 of one kind, 2 of another kind, and 2 of yet another kind. Answer: $\binom{13}{3}\binom{4}{2}^3$

7. Count the number of hands from the deck that consist of 2 of one kind, 2 of another kind, and 3 of yet another kind. Answer: $11 \cdot 4 \cdot \binom{13}{2}\binom{4}{2}^2$

8. Count the number of five card hands from the deck that are straights. That is, they contain 5 consecutive symbols. For example, the hand might contain $2, 3, 4, 5, 6$ where any suits are allowed for these numbers. Answer: $9 \cdot 4^5$

Chapter 8

Arrangement Problems

The problems we have discussed to date are mostly problems in which the symbols do not repeat. The larger problem broke down into a simple choice of one element from a larger set, or a permutation problem, or a combination problem. Our first problem will lead us past the last few chapters.

As we practiced before, words need not be in a dictionary. They are just strings of letters. An *arrangement of the word* W is a word that uses the same letters that appear in W and with the same frequency. For example, an arrangement of

$$A^3B^4C^5 = AAABBBBCCCCC$$

will be a word that consists of 3 A's, 4 B's, 5 C's, and nothing more. As another example, CAB is an arrangement of ABC, while $AABC$ and ABD are not.

The words we will consider can have repeated letters so this is not a permutation problem. Permutation problems require that the word has no repeated symbols. Thus, we need a new technique if we are to solve this problem.

Consider the problem of constructing arrangements. An arrangement of a word is completely determined by where each letter is placed. If I say that A appears in the first, the fifth, and the last place of the arrangement, then you know exactly where A occurs. If I say that A is in places 2, 4, and 6 then you know where all of the A's are in that arrangement. Thus, to construct an arrangement, you specify which places contain the A's, which places contain the B's, and which places contain the C's. If there are more letters, you continue. If there are 3 A's in the word then you place 3 A's. If there are 7 D's in the word then you place 7 D's in the arrangement. You continue until you run out of letters.

8.1 Examples of Arrangements

EXAMPLE 8.1.1 Count the arrangements of the word

$$A^3B^4C^5 = AAABBBBCCCCC.$$

There are 12 places in this word since there are 12 letters in this word. To form an arrangement, we place letters. There are 3 A's so our first task is to choose the 3 places that will contain A. We choose 3 of 12 places to hold the A's. This is a combination problem which can be done in $\binom{12}{3}$ ways to do this.

Nine places remain. The next task is to place the 4 B's. Choose the 4 of 9 remaining places to contain B. This is a combination problem done in $\binom{9}{4}$ ways.

The last task is to place the 5 C's in the remaining 5 places. This is done in 1 way.

Then this sequence of tasks can be done in

$$\binom{12}{3}\binom{9}{4}$$

ways.

EXAMPLE 8.1.2 We could just as easily have formed our arrangement of $A^3B^4C^5$ by first choosing 5 of 12 places for the C's. This is done in $\binom{12}{5}$ ways. Choose the 4 places of the remaining 8 places to hold the B's. This is done in $\binom{7}{4}$ ways. There are just enough places to hold the 3 A's and there are only 3 places left, so we choose all of the 3 places in 1 way.

Then the number of arrangements in our word $A^3B^4C^5$ is

$$\binom{12}{5}\binom{7}{4}$$

ways. A quick calculation shows that this is the same number we found in the previous example. Of course it is. We are counting the same objects.

EXAMPLE 8.1.3 The classic counting problem at this time is to find the number of arrangements of $MISSISSIPPI$.

Observe that $MISSISSIPPI$ is an arrangement of $MI^4S^4P^2$. There are 11 places in this word. Choose 1 place of 11 for the M. This is done in 11 ways. Next choose 4 places from the remaining 10 for the I. This is done in $\binom{10}{4}$ ways. Choose 4 of 6 places to hold the S's. This is done in $\binom{6}{4}$ ways. There is only 1 way to choose the remaining 2 places for the 2 P's. Then the number of arrangements of $MI^4S^4P^2$ is

$$11\cdot\binom{10}{4}\binom{6}{4}.$$

EXAMPLE 8.1.4 Count the number of 10 letter words that contain exactly 5 a's, exactly 3 b's, and 2 other letters (perhaps repeated).

The word has 10 places in it. Choose 5 of 10 places for the a. This is a combination problem so it is done in $\binom{10}{5}$ ways. Of the remaining 5 places, choose 3 for the b's. This is a combination problem and is done in $\binom{5}{3}$ ways. Two places remain to be filled with letters other than a and b. Since each place can be filled with any one of the 24 remaining letters, the remaining two places can be filled in 24^2 ways. Therefore, the total number of words containing exactly 5 a's and 3 b's is

$$\binom{10}{5}\binom{5}{3}24^2.$$

Let us solve the general problem on arrangements. We need a little terminology. In Example 8.1.1, A *appears* 3 times in $W = A^3B^4C^5$, B *appears* 4 times in W, and C *appears* 5 times in W.

EXAMPLE 8.1.5 Let $W = L_1^{m_1}L_2^{m_2}L_3^{m_3}$ be a word with 3 different letters L_1, L_2, and L_3. By our exponent notation, L_1 appears m_1, L_2 appears m_2, and L_3 appears m_3 times in W. We count the number of arrangements of W.

Since the letters L_1, L_2, and L_3 are the only letters that are used to form W, the number of letters in W is

$$M = m_1 + m_2 + m_3.$$

Then W has exactly M places into which we will place the letters L_1, L_2, and L_3.

Task 1 is to choose m_1 places from M that will hold L_1. This can be done in $\binom{M}{m_1}$ ways.

There remain $M - m_1$ places. Task 2 is to choose m_2 places from the remaining $M - m_1$ that will hold L_2. This is done in $\binom{M - m_1}{m_2}$ ways.

Since $M = m_1 + m_2 + m_3$ there remains $M - m_1 - m_2 = m_3$ places to hold the letter L_3. There is only 1 way to choose m_3 places for L_3 from the remaining m_3 places.

Thus,

$$\text{the number arrangements of } W = L_1^{m_1}L_2^{m_2}L_3^{m_3} = \binom{M}{m_1}\binom{M - m_1}{m_2}.$$

Suppose we write out the combination numbers in the previous example as fractions and then simplify the expression. There results equations

$$\begin{aligned}\binom{M}{m_1}\binom{M - m_1}{m_2} &= \frac{M!}{m_1!(M - m_1)!}\frac{(M - m_1)!}{m_2!(M - m_1 - m_2)!} \\ &= \frac{M!}{m_1!}\frac{1}{m_2!(M - m_1 - m_2)!}.\end{aligned}$$

Since $M = m_1 + m_2 + m_3$, we have $M - m_1 - m_2 = m_3$, so that

$$\begin{aligned}\text{the number of arrangements of } W &= L_1^{m_1} L_2^{m_2} L_3^{m_3} \\ &= \frac{M!}{m_1!m_2!m_3!}.\end{aligned}$$

An application of this formula is to count the arrangements of the word

$$W = 1^4 3^4 5^4 7^4 = 1111333355557777.$$

You say this is a number, but it is a word in this problem. W consists of 16 digits. The digits in W are 1, 3, 5, and 7 and each appears exactly 4 times. Thus, the number of arrangements of W is

$$\binom{16}{4}\binom{12}{4}\binom{8}{4} = \frac{16}{(4!)^3}.$$

We will now solve the general problem and see that an interesting formula occurs.

EXAMPLE 8.1.6 Let $W = L_1^{m_1} L_2^{m_2} \cdots L_t^{m_t}$ be a word with symbols $L_1, L_2, \ldots, L_t$. By our notation, L_1 appears exactly m_1 times, L_2 appears m_2 times, ..., and L_t appears m_t times in W. The number of places in W is

$$M = m_1 + m_2 + \cdots + m_t.$$

By choosing m_1 of M places for L_1, m_2 places of $M - m_1$ for L_2, and so on (these are the same tasks that we used in the previous examples, but using them more often), we arrive at the formula

$$\binom{M}{m_1}\binom{M - m_1}{m_2}\cdots\binom{M - m_1 - m_2 - \cdots - m_{t-1}}{m_t}$$

for the number of arrangements of W. This will simplify significantly.

Take any two of the combination numbers in a row. Say we take combination number i and the factor that comes after it.

$$\binom{M - m_1 - m_2 - \cdots - m_{i-1}}{m_i}\binom{M - m_1 - m_2 - \cdots - m_i}{m_{i+1}}$$

After writing these numbers as fractions we have

$$\frac{(M - m_1 - m_2 - \cdots - m_{i-1})!}{m_i!(M - m_1 - m_2 - \cdots - m_i)!}\frac{(M - m_1 - m_2 - \cdots - m_i)!}{m_{i+1}!(M - m_1 - m_2 - \cdots - m_{i+1})!}.$$

Observe that there is a cancellation that occurs in these fractions. The reduced fractions are

$$\frac{(M - m_1 - m_2 - \cdots - m_{i-1})!}{m_i!}\frac{1}{m_{i+1}!(M - m_1 - m_2 - \cdots - m_{i+1})!}.$$

Then the complete product

$$\frac{M!}{m_1!(M - m_1)!}\frac{(M - m_1)!}{m_2!(M - m_1 - m_2)!}\frac{(M - m_1 - m_2)!}{m_3!(M - m_1 - m_2 - m_3)!}\cdots$$

reduces to

$$\frac{M!}{m_1!}\frac{1}{m_2!}\frac{1}{m_3!}\cdots.$$

The cancellation continues until we reach the t-th factor. The end of the factors is

$$\cdots\frac{(M - m_1 - m_2 - \cdots - m_{t-1})!}{m_t!(M - m_1 - m_2 - \cdots - m_{t-1} - m_t)!}.$$

Since $M = m_1 + m_2 + \cdots + m_t$ we have $M - m_1 - m_2 - \cdots - m_{t-1} = m_t$ and $M - m_1 - m_2 - \cdots - m_t = 0$. Our work above shows that the factor $(M - m_1 - m_2 - \cdots - m_{t-1})!$ cancels with a copy of $(M - m_1 - m_2 - \cdots - m_{t-1})!$ in the denominator of the previous fraction. Thus,

$$\cdots\frac{(M - m_1 - m_2 - \cdots - m_{t-1})!}{m_t!(M - m_1 - m_2 - \cdots - m_{t-1} - m_t)!}$$
$$= \cdots\frac{1}{m_t!0!} = \cdots\frac{1}{m_t!}.$$

Therefore,

$$\text{the number of arrangements of } W = L_1^{m_1} L_2^{m_2} \cdots L_t^{m_t} = \frac{M!}{m_1! m_2! \cdots m_t!}.$$

This number is called a *multinomial coefficient.*

A few examples might help you understand how to use the formula.

EXAMPLE 8.1.7 Suppose we must count the arrangements of

$$W = K^4 L^5 M^4 N^5 = KKKKLLLLLMMMMNNNNN.$$

W has 18 places so the number of arrangements of W is

$$\frac{18!}{4!4!5!5!} = \frac{18!}{(4!)^2(5!)^2}.$$

EXAMPLE 8.1.8 Count the number of arrangements of $W = A^{10} B^{10} \cdots Z^{10}$ where the letters range over the capital letters in the English alphabet. Each letter is repeated 10 times in W, and the number of places in W is 260. Thus, the number of arrangements of W is

$$\frac{260!}{(10!)^2 6}.$$

EXAMPLE 8.1.9 Let us see how this formula works in a minimal situation, that is, when all of the letters appear just once. The arrangements of the word $W = ABCDEF$ is

$$\frac{6!}{1!1!1!1!1!1!} = 6!$$

which is what is predicted by the more elementary method of a permutation number.

8.2 Exercises

Use the formulas in this to solve these problems. Let $W = 9^6 3^6 A^5 B^5$.

1. Count the number of arrangements of W using the method outlined in Example 8.1.1. Answer: $\binom{22}{6}\binom{16}{6}\binom{10}{5}$

2. Count the number of arrangements of W using a multinomial coefficient. Answer: $\dfrac{22!}{22!16!10!}$

3. Let $V = a_1 a_2 \cdots a_n$ be a word without repeated symbols. Count the number of arrangements by using the methods outlined in this chapter and then by using the methods used to count permutations. Show that these numbers are the same.

4. Suppose that V is a word with symbols A, B, C, and D. Suppose that A appears 10 times in W, that B appears 20 times in W, that C appears 30 times in W, and that D appears 40 times in W. Using the argument given in Example 8.1.1, find the number of arrangements of V. Answer: $\binom{100}{10}\binom{90}{20}\binom{70}{30}$

5. Simplify the answer found in exercise 4, and show that it is equal to $\dfrac{100!}{10!20!30!40!}$.

Chapter 9

At Least, At Most, and Or

Most problems cannot be solved with just one counting formula. They require an indepth examination that leads us to, once more, smaller more manageable problems. To get to these smaller problems we need to understand how the language of the larger problem suggests a solution. In this chapter we will investigate how phrases such as *at least*, *at most*, and *or* allow us to analyze the given problem. We will find that some language such as *or* corresponds in a one-to-one fashion with addition, while *at least* and *at most* will often lead to a series of smaller problems that we have already solved.

9.1 Counting with *Or*

Let T be a task. An *outcome of* T is what results when we enact T. So if T is the task to choose a card from a deck, then K is one of the outcomes of T, as is Q. If T is the task of flipping a coin, then an outcome of T is to observe a Head or H.

Two tasks S and T are said to be *disjoint* if no object is an outcome of both S and T. Thus, if x is an outcome of the task S (if x is formed by the process S), then x is not an outcome of the task T. Symmetrically, no outcome of T is an outcome of S. Here is another way to understand disjointness. Let X be the set of outcomes of S and let Y be the set of outcomes of T. Then S and T are disjoint provided that $X \cap Y = \emptyset$.

If T is a task then let

$$n(T) = \text{the number of outcomes of } T.$$

Thus, if T is the task of choosing a letter from the English alphabet, $n(T) = 26$. If T is the task of choosing an even number from the set of digits then $n(T) = 5$ since the outcomes are $\{0, 2, 4, 6, 8\}$. If T is the task of making a permutation of 37 symbols from an alphabet of 177 different symbols then $n(T) = P(177, 37)$. If T is the task of choosing 75 elements from a set of 125 elements then $n(T) = \binom{125}{75}$.

The disjoint tasks are the ones we need to make our counting easy. The following formula shows us why this is True.

Let S and T be disjoint tasks. Let S and T be tasks, let X be the set of outcomes of S, and let Y be the set of outcomes of T. Then by the inclusion/exclusion principle,

$$n(X \cup Y) = n(X) + n(Y) - n(X \cap Y).$$

Since S and T are disjoint, $X \cap Y = \emptyset$ so that $n(X \cap Y) = n(\emptyset) = 0$. Thus,

$$n(X \cup Y) = n(X) + n(Y).$$

Now $X \cup Y$ is the set of outcomes of S or T, so we can write $n(S \text{ or } T) = n(X \cup Y)$. We have derived a very important formula in our counting explorations.

If S and T are disjoint tasks, then

$$n(S \text{ or } T) = n(S) + n(T)$$

Here are some examples of how to use our new tool.

EXAMPLE 9.1.1 Count the number of ways for a 3-letter word to contain exactly 1 a or exactly 2 a's.

There are two tasks S and T at work here. The task S is to choose a 3-letter word with exactly 1 a. If a word contains exactly one a then choose the 1 of the 3 places for the a. This is done in 3 ways. The two remaining places can contain any letter but a. Each place can be filled with 1 of 25 letters. Thus. the number of ways to accomplish S is $n(S) = 3 \cdot 25^2$.

The task T is to choose a 3-letter word containing exactly 2 a's. Choose the 2 places in 3 to contain the 2 a's in $\binom{3}{2} = 3$ ways. Since this word contains exactly 2 a's, we fill the remaining place with 1 of 25 letters. This is done in 25 ways. The number of ways to accomplish T is $n(T) = 25 \cdot 3$ ways.

Therefore, the total number of words that contain exactly 1 or exactly 2 a's is

$$3 \cdot 25^2 + 25 \cdot 3 = 3 \cdot 25 \cdot 26.$$

EXAMPLE 9.1.2 Count the number of 4-letter words that contain exactly 3 x's or exactly 4 x's.

Solution: We break the problem down into two smaller problems. One smaller problem is to count the 4 letter words that contain exactly 4 x's. There is only 1 such word.

The other smaller problem is to count the 4-letter words containing exactly 3 x's. To do this, choose the 3 places of 4 to contain the x's. This is done in $\binom{4}{3} = 4$ ways. Then choose a letter, not an x, for the remaining place. This is done in 25 ways. This task is done in $4 \cdot 25$ ways.

Then the number of 45-letter words containing exactly 3 x's or exactly 4 x's is

$$1 + 4 \cdot 25 = 101.$$

Students sometimes try to count these *at least* problems with an *erroneous* counting technique. Let me demonstrate. We will make a count of the number of small words from a small alphabet in three ways. From this discussion, we will see which counting technique is correct and which is in error.

EXAMPLE 9.1.3 We will count the number of 2-letter words from the alphabet $\{a, b, c\}$ that contain at least 1 a. We will do this in the manner that we have established to this point. Then we will count using a commonly used and incorrect technique. Then we will make a list of all of these words and compare the numbers.

1. To count the 2-letter words from $\{a, b, c\}$ that contain at least 1 a, count the number of words with exact 1 a, and the number of such words with exactly 2 a's.

A word with exactly 1 a is constructed by first placing the a in one of the 2 places in 2 ways. Then, because there are exactly 2 a's in this case, you fill the other place with one of the letters that is not an a in 2 ways. The total is $n(\text{exactly } 1\ a) = 2 \cdot 2 = 4$.

There is exactly 1 two-letter word with exactly 2 a's. It is aa. So $n(\text{exactly } 2\ a\text{'s}) = 1$.

This technique shows that the number of 2-letter words with at least 1 a is exactly $4 + 1 = 5$.

2. Let us try that common flawed technique. In this technique, you place an a in one of two places in 2 ways, and then you fill in the remaining place with one of the 3-letters in the alphabet $\{a, b, c\}$. This second selection is made in 3 ways. According to this technique the total number of 2-letter words with at least 1 a is $2 \cdot 3 = 6$.

3. So which is it? We arrived at numbers 5 and 6 for the number of 2-letter words from $\{a, b, c\}$ with at least 1 a. One of the numbers is incorrect, but which one? The only way to know which is make a list of all of the 2-letter words from $\{a, b, c\}$ with at least 1 a. That list is formed by placing the letter a in the first place and the second place, and then by filling in the empty place. Observe.

$$\begin{array}{ccc} aa & ab & ac \\ aa & ba & ca \end{array}$$

It appears that we have 6 words until you examine the first word, aa. It is listed twice and therefore it is counted twice. This double counting makes an error in our counting. That is why the method in item 2 is in error. It counts aa twice. The correct list of our words is found by deleting one copy of aa from the first list, as follows.

$$\begin{array}{ccc} aa & ab & ac \\ & ba & ca \end{array}$$

There are just 5 words now. Each is listed exactly once. We do not count a word twice. Thus, the number of 2-letter words from $\{a, b, c\}$ with at least 1 a is correctly given as 5. This means that the method in item 1 produces the correct answer.

To those who don't understand why counting aa twice in the above example leads to an incorrect answer, let us use money. You have a pile of 100 dollar bills and you are trying to count them. You begin by counting the first bill *twice*. You have just counted 101 dollars, when in fact only 100 dollars is in your hand. So which technique is correct? Obviously, the technique that counts exactly 100 dollars is correct. Thus, when counting things *it is important that we count things only once.*

EXAMPLE 9.1.4 A larger problem is in order. We will count the number of 3-letter words from $\{a, \ldots, z\}$ containing at least 1 a. We will use the same approach that we used in Example 9.1.3.

1. To count at least 1 a, we will count exactly 1 a, exactly 2 a's, and exactly 3 a's.

Words with exactly 1 a are constructed by placing 1 a in one of the 3 places in 3 ways. Then because these words have exactly 1 a, you fill in the 2 remaining places with letters from $\{b, c, \ldots, z\}$. You fill each empty place in the word with one of 25 letters in 25^2 ways.

Words with exactly 2 a's are constructed by placing 2 a's in 2 of the 3 places. This is done in exactly $\binom{3}{2}$ ways. The third place cannot contain an a since these words have exactly 2 a's. You fill in the remaining place with one of the letters $\{b, c, \ldots, z\}$ in 25 ways.

There is only 1 three-letter word with exactly 3 a's. It is aaa.

Thus, the total number of 3-letter words containing at least 1 a is $3 \cdot 25^2 + \binom{3}{2} \cdot 25 + 1 = 3 \cdot 25^2 + 3 \cdot 25 + 1$.

2. Let use try the common but incorrect technique for counting words with at least 1 a. Proceed by placing a in 1 of 3 places in 3 ways. Then fill in the other 2 places with one of 26 letters in 26^2 ways. This technique gives us $3 \cdot 26^2$ words.

3. Two different numbers are found. You can determine that these are unequal numbers by using a calculator, or by noting that $3 \cdot 25^2 + 3 \cdot 25 + 1$ is no divisible by 3 while $3 \cdot 26^2$ is divisible by 3.

Which technique correctly counts the number of 3-letter words with at least 1 a? We answer the question by making a complete list of the words to be counted. I think we can all agree that a 3-letter word has no a's, exactly 1 a, exactly 2 a's, or exactly 3 a's. We can ignore words with no a's because the words we are counting have at least 1 a.

Make a list of the words with exactly 1 a.

$$aXY \quad XaY \quad XYa$$

where X and Y are letters. Because there is exactly 1 a in these words X and Y come from the 25 letter alphabet $\{b, c, \ldots, z\}$.

Next, make a list of the words with exactly 2 a's.

$$aaX \quad aXa \quad Xaa$$

where X is a letter. Because there are exactly 2 a's in these words, X comes from the alphabet $\{b, c, \ldots, z\}$ consisting of 25 letters.

Finally, there is only 1 word aaa containing exactly 3 a's.

The total number of words counted in this list is $25^2 + 25^2 + 25^2 + 25 + 25 + 25 + 1 = 3 \cdot 25^2 + 3 \cdot 25 + 1$. The first counting technique again gives the correct answer.

4. Let us see why item 2 does not produce the correct number of words. Examine the list of 3-letter words produced by the technique. The letter a is placed in the word and the other letters fill in the remaining places. The result is a list

$$aXY \quad XaY \quad XYa$$

where X, $Y \in \{a, \ldots, z\}$. Now *here is the error.* One possible value for X and Y is a. But when you allow $X = Y = a$ then you have counted the word aaa once as aXY, once as XaY, and once as XYa. We have counted aaa three times. This is, of course, why the number in item 2 is incorrect.

EXAMPLE 9.1.5 Count the number of 4-letter words whose letters are from $\{a, \ldots, z\}$ and that contain at least 3 a's. Since you want words containing exactly 3 a's or 4 a's. Choose 3 of 4 places for the a's in $\binom{4}{3} = 4$ ways. Then *erroneously* fill in the remaining place with any one of 26 letters including the a. This is done in 26 ways. The number produced is $4 \cdot 26$, but it is not a count of the 4-letter words with at least 3 a's. The reason the number is in error is that some words are being counted more than once. Observe.

Place the 3 a's to begin to form a word. Here are two such placements of the a's. There are several more that we will not need.

$$aaaX \text{ and } Xaaa$$

The words that we can form are made by filling in the blank space with a letter X. When you count these words, you count them as two different words, *and that is the error.* Look at what happens when you let $X = a$. The result is the one word

$$aaaa \text{ and } aaaa$$

formed in two different ways. We have already observed that counting a word twice yields an inccorect accounting of the words in our problem. **Moral:** *Be very careful when looking for shortcuts in these problems.*

9.2 At Least, At Most

The term *at least* n means exactly n, or exactly $n + 1$, or exactly $n + 2, \ldots$. You stop when the problem forces you to stop. If you want to count the 5 words with *at least 3* vowels, then you will analyze words with exactly 3 vowels, or exactly 4 vowels, or exactly 5 vowels. If you want to count the number of sets with at least 7 digits, you would analyze the problem by counting sets with exactly 7 digits, or exactly 8 digits, or exactly 9 digits, or exactly 10 digits. You would stop here because there are only 10 digits.

The *or* used shows us that the following suggestive formula can be used. Suppose you have a task that can be accomplished in as many as n ways. Then

counting *at least k ways* is the same as counting exactly k ways or exactly $k+1$ ways or $\cdots$ or exactly n ways.

Since the individual tasks *exactly k ways* and *exactly $k+1$ ways* cannot occur at the same time, these tasks are disjoint tasks. Thus, by the formula on page 100, these *or* statements translate into addition when we count.

$$n(\text{at least } k \text{ ways}) = n(\text{exactly } k \text{ ways}) + n(\text{exactly } k+1 \text{ ways}) + \cdots + n(\text{exactly } n \text{ ways})$$

The notation is clear. n(exactly 3 ways) means that this is the number of outcomes having exactly 3. What 3 things will depend upon the problem being worked. It might be repeated letters, or it might be open spaces, but we will count the exact number of times 3 ways are observed. Thus, a statement counting at least k ways will be translated into a series of smaller problems, and these are conjuncted with an *or*.

EXAMPLE 9.2.1 Let T be some task that can be done in up to 5 ways. Then the number of ways that T can be done in at least 3 ways is counted as follows:

$$n(\text{at least 3 ways}) = n(\text{exactly 3 ways}) + n(\text{exactly 4 ways}) + n(\text{exactly 5 ways})$$

Of course, now we would have to examine and count the numbers n(exactly k ways) for $k = 3, 4, 5$.

EXAMPLE 9.2.2 Count the number of 6 letter words that contain at least 5 repeated a's.

We analyze the count. First we examine the defining terms *at least 5 a's*.

at least 5 a's means *exactly 5 a's* or *exactly 6 a's*.

There are only 6 letters in the word, so we stop at 6 a's. Then counting yields

$$n(\text{at least } 5\ a\text{'s}) = n(\text{exactly } 5\ a\text{'s}) + n(\text{exactly } 6\ a\text{'s}).$$

Of course, now we have to count the numbers $n(\text{exactly } 5\ a\text{'s})$ and $n(\text{exactly } 6\ a\text{'s})$.

To calculate $n(\text{exactly } 5\ a\text{'s})$ we choose the 5 of 6 places to hold the a's. This is done in 6 ways. Fill the remaining place with a letter not equal to a is 25 ways. Then $n(\text{exactly } 5\ a\text{'s}) = 25 \cdot 6$.

Observe that $n(\text{exactly } 6\ a\text{'s}) = 1$ since there are only 6 places to fill.

Then $n(\text{at least } 5\ a\text{'s}) = n(\text{exactly } 5\ a\text{'s}) + n(\text{exactly } 6\ a\text{'s}) = 25 \cdot 6 + 1$.

EXAMPLE 9.2.3 Analyze the number of 4 digit serial numbers (created from the set of digits) that contain at least 2 repeated 1's.

In this case, *at least two* 1*'s* means

exactly two 1*'s* or *exactly three* 1*'s* or *exactly four* 1*'s.*

There are only 4 digits in the word, so we stop at *exactly 4* 1*'s.* Then counting yields $n(\textit{at least } 2\ 1\textit{'s}) =$

$$n(\textit{exactly two } 1\textit{'s}) + n(\textit{exactly three } 1\textit{'s}) + n(\textit{exactly four } 1\textit{'s}).$$

We count the numbers $n(\textit{exactly } k\ 1\textit{'s})$ for $k = 2, 3, 4$. We start with $k = 4$.

The number of 4-digit serial numbers containing exactly four 1's is 1.

$$n(\textit{exactly } 4\ 1\textit{'s}) = 1.$$

The number of 4-digit serial numbers containing exactly three 1's is counted by choosing the 3 of 4 places to fill with a 1. This is done in 4 ways. Next fill the 4-th place with one of 9 digits (not 1) in 9 ways. Then

$$n(\textit{exactly } 3\ 1\textit{'s}) = 4 \cdot 9.$$

The number of 4-digit serial numbers containing *exactly* two 1's is counted by choosing the 2 of 4 places to fill with a 1. This is done in $\binom{4}{2}$ ways. Next fill the remaining 2 places with one of 9 digits (not 1) each. This is done in 9^2 ways. Then

$$n(\text{exactly two 1's}) = \binom{4}{2} \cdot 9^2$$

ways.

Then

$$n(\text{at least two 1's}) = 1 + 4 \cdot 9 + \binom{4}{2} \cdot 9^2.$$

The term *at most k* means exactly 0, or exactly 1, or ..., or exactly k. The counting begins with 0 and ends with k. Thus, when asked to count 7-letter words with at most 2 x's you would count words containing exactly 0 x's, words containing exactly 1 x, and words containing exactly 2 x's. These are disjoint tasks. The formula on page 100 counting disjoint tasks shows us that

$$n(\text{at most } k \text{ ways}) = n(\text{exactly 0 ways}) + n(\text{exactly 1 way}) + \cdots + n(\text{exactly } k \text{ ways}).$$

EXAMPLE 9.2.4 Count the number of 10-letter words in the English language that contain at most 2 a's.

At most two ways means *exactly 0 ways* or *exactly 1 ways* or *exactly 2 ways.* To count such a problem we will translate *or* into + as follows.

$$n(\text{at most two } a\text{'s}) = n(\text{exactly 0 } a\text{'s}) + n(\text{exactly 1 } a\text{'s}) + n(\text{exactly 2 } a\text{'s}).$$

We must count each term in this sum.

Exactly 0 a's means that the 10 letters used are from the alphabet excluding a. There are 25 letters in that alphabet and 10 places to fill. There are 25^{10} such words.

Exactly 1 a's Choose 1 of 10 places for the a in 10 ways. The remaining 9 places are filled with letters from the alphabet excluding a. This alphabet has 25 letters. Thus, there are 25^9 ways to fill the remaining 9 places. Then $n(\textit{exactly 1 a}) = 10 \cdot 25^9$.

Exactly 2 a's means that there are exactly 2 a's in the word, and the other 8 letters are different from a. Choose 2 of 10 places to fill with a in $\binom{10}{2}$ ways. As in the above two cases, the other 8 places are filled in 25^8 ways. Thus, $n(\textit{exactly 2 a's}) = \binom{10}{2} \cdot 25^8$.

Hence

$$n(\text{at most 2 } a\text{'s}) = 25^{10} + 10 \cdot 25^9 + \binom{10}{2} \cdot 25^8.$$

9.3 Exercises

The words *word* in these exercises refers simply to a string of the English alphabet.

1. Count the number of 7-letter words that either contain exactly 3 b's or that begin with a vowel. (Count y as a vowel here.) Answer: $\binom{7}{3}25^4 + \binom{7}{4}25^3 + \binom{7}{5}25^2 + 7 \cdot 25 + 1 + 6 \cdot 20^6 - 6 \cdot \left(\binom{6}{3}25^3 + \binom{6}{4}25^2 + 6 \cdot 25 + 1\right)$.

2. Count the number of 8-letter words that contain one letter repeated at least 6 times. Answer: $26 \cdot \binom{8}{6} \cdot 25^2 + 26 \cdot 8 \cdot 25 + 26$

3. Count the number of 12-letter words that contain one letter repeated at most 3 times. The letter must appear in this word. Answer: $26 \cdot \binom{12}{3} \cdot 25^9 + 26 \cdot \binom{12}{2} \cdot 25^{10} + 26 \cdot 12 \cdot 25^{11}$

4. Count the number of 15-letter words that contain either one letter repeated at least 14 times or one letter repeated at most 2

times. Answer: $26 \cdot 15 \cdot 25 + 26 \cdot 25 + 26 \cdot \binom{15}{2} \cdot 25^{13} + 26 \cdot 15 \cdot 25^{14}$

5. Count the number of 20-letter words that contain one letter repeated at most 2 times or one letter repeated at least 19 times. Answer: $26 \cdot \binom{20}{2} \cdot 25^{18} + 26 \cdot 20 \cdot 25^{19} + 26 \cdot 20 \cdot 25 + 26$

Chapter 10

Complement Counting

There are a few words that make for complicated counting problems and that at first sight are solved with a lot of writing. The few words I am referring to are

at least one.

When a problem uses a task that can be done in n ways, and when we are asked to count the number of ways that this task can be completed in *at least 1 way*, we are presented with a chain of at least $n-1$ problems to solve. We have solved that problem in the previous chapter when we write that *at least 1 way* means *exactly 1 way* or *exactly 2 ways* or ... or *exactly n ways*. This solution to *at least 1 way* will usually require a lot of writing. In this chapter we will look at solutions to the *at least 1 way* counting problem that can be reduced to a couple of manageable smaller counting problems.

10.1 The Complement Formula

Recall that if A is a set and if $\mathcal{U}$ is a universal set containing A then

$$A' = \{x \in \mathcal{U} \mid x \notin A\}.$$

The set A' is just the set of elements in $\mathcal{U}$ that are *not* in A. See Section 2.2 for details. Consider an element $x \in \mathcal{U}$. Since A is a set, either $x \in A$ or $x \notin A$. That is, x is in either A or in A'. This shows us that

$$\mathcal{U} = A \cup A'.$$

We can even count $\mathcal{U}$ using the inclusion/exclusion principle with this union. Observe.

$$\begin{aligned} n(\mathcal{U}) &= n(A \cup A') \\ &= n(A) + n(A') - n(A \cap A') \\ &= n(A) + n(A') - 0 \text{ because } A \cap A' = 0 \\ &= n(A) + n(A') \end{aligned}$$

Then $n(\mathcal{U}) = n(A) + n(A')$ so that $n(\mathcal{U}) - n(A') = n(A)$. We have thus established the important complement formula.

$$n(A) = n(\mathcal{U}) - n(A')$$

Our examples of how to use the complement formula will start slowly and build.

EXAMPLE 10.1.1 Count the number of letters in the English alphabet that are different from a. Of course there are 25 letters different from a, but this is an elementary example that I will use to make a point.

Let $A =$ the set of letters that are not a. This is what we want to count. The universal set for this problem is $\mathcal{U} =$ the alphabet. It contains 26 elements. Then $A' = \{a\} =$ the set of letters that are a. It contains 1 element. Then by the complement formula

$$n(A) = n(\mathcal{U}) - n(A') = 26 - 1 = 25.$$

EXAMPLE 10.1.2 Count the number of 3-letter words that contain an a.

We solve the problem using the complement formula. $A =$ the set of 3-letter words that contain an a. $\mathcal{U} =$ the set of all 3-letter words. $A' =$ the set of 3-letter words that do not contain an a. Then $n(\mathcal{U}) = 26^3$ because we must fill 3 places with one of 26 letters. Furthermore, count A'. The words in A' consist of 3 letters other than a. There are no other restrictions, so each place is filled with 1 of 25 letters. Then by the multiplication principle $n(A') = 25^3$, so by the complement formula

$$n(A) = n(\mathcal{U}) - n(A') = 26^3 - 25^3.$$

EXAMPLE 10.1.3 Count the number of 10-digit serial numbers that do not begin with 0.

Let $A =$ the set of 10-digit serial numbers that do not begin with 0. $\mathcal{U} =$ the set of all possible 10-digit serial numbers. $A' =$ the set of 10-digit serial numbers that begin with 0. We have counted $\mathcal{U}$ a number of times. $n(\mathcal{U}) = 10^{10}$. To count A', note that the serial numbers begin with 0, so there is only one choice for the first place in a word in A'. The other 9 places can be filled with 10 digits each, so by the multiplication principle $n(A') = 9 \cdot 10^9$. Hence,

$$n(A) = n(\mathcal{U}) - n(A') = 10^{10} - 9 \cdot 10^9.$$

EXAMPLE 10.1.4 Count the number of subsets of $\{0, 1, 2, 3, 4, 5, 6, 7, 8, 9\}$ that do not contain $\{2, 4, 6, 8\}$.

The universal set in this problem is $\mathcal{U} =$ the set of all subsets of $\{0, 1, 2, 3, 4, 5, 6, 7, 8, 9\}$. Then $n(\mathcal{U}) = 2^{10}$.

Let A be the set of all subsets of $\{0, 1, 2, 3, 4, 5, 6, 7, 8, 9\}$ that do not contain the set $\{2, 4, 6, 8\}$.

A' is the set of subsets of $\{0, 1, 2, 3, 4, 5, 6, 7, 8, 9\}$ that contain $\{2, 4, 6, 8\}$. Consider for a moment. $X = \{1, 2, 4, 6, 8, 9\}$ is completely determined by the set $\{2, 4, 6, 8\}$ and the set $\{1, 9\}$. $X = \{2, 3, 4, 5, 6, 7, 8\}$ is completely determined by $\{2, 4, 6, 8\}$ and the set $\{3, 5, 7\}$. $X = \{2, 4, 6, 8\}$ is completely determined by $\{2, 4, 6, 8\}$ and the set $\emptyset$. Thus, a subset X of $\{0, 1, 2, 3, 4, 5, 6, 7, 8, 9\}$ that contains $\{2, 4, 6, 8\}$ is completely determined by the largest subset of $\{0, 1, 3, 5, 7, 9\}$ that X contains. There are 2^6 subsets of $\{0, 1, 3, 5, 7, 9\}$. Thus, there are 2^6 subset X of $\{0, 1, 2, 3, 4, 5, 6, 7, 8, 9\}$ that contain $\{2, 4, 6, 8\}$. That is, $n(A') = 2^6$.

Hence,

$$n(A) = n(\mathcal{U}) - n(A') = 2^{10} - 2^6.$$

10.2 A New View of "At Least"

We will show how to use the complement formula to reduce the amount of writing needed to solve certain *at least 1 way* problems.

Suppose we are given a task T that can be done in n ways, and suppose we are to count the number of ways that T can be done in *at least 1 way*. Let $\mathcal{U}$ be the set of ways that T can be done, and let A be the set of ways that T can be done in at least 1 way. Then

$$A' = \text{the set of ways that } T \text{ can be done in} \\ \textit{exactly 1 way} \text{ or } \textit{done not at all.}$$

Then

$$n(A') = n(\text{exactly 1 way}) + n(\text{done not at all}).$$

Ordinarily, n(done not at all) is not a hard problem that can be found quickly. So the real problem in counting A' is finding n(exactly 1 way). We have solved a number of problems like this. So to find $n(A)$, we have reduced the problem to one of finding the number of ways T can be done exactly once. Some examples will help illuminate the method.

EXAMPLE 10.2.1 Count the 10 letter words from the English alphabet that have at least 1 a.

We can solve the problem directly but to demonstrate the new method, we will solve this problem indirectly.

$\mathcal{U}$ = the set of all 10-letter words from the English alphabet, and let A = the set of all 10-letter words that contain at least 1 a. Then A' = is the set of all 10-letter words that do not contain an a.

We have counted the number of 10-letter words in Example 4.1.8. $n(\mathcal{U}) = 26^{10}$.

Count A'. Ten-letter words without an a are 10-letter words from the 25 letter alphabet $\{b, c, \ldots, x, y, z\}$. We have counted these words before. $n(A') = 25^{10}$.

Therefore,

$$n(A) = 26^{10} - 25^{10}.$$

In the following problems about chips in a bag, we have a bag of 100 chips colored red, white, blue, and black. The red colored chips are numbered 1 to 25, the white colored chips are numbered 1 to 25, the blue colored chips are numbered 1 to 25, and the black colored chips are numbered 1 to 25.

EXAMPLE 10.2.2 We choose a handful of 40 chips from the bag.

1. Count the number of handfuls that contain at least 1 red chip.

2. Count the number of handfuls that contain at least 2 red chips.

1. Let A be the set of handfuls with at least 1 red chip. We count the complement. Then A' = the set of handfuls of 40 chips with no red chips.

To count A', choose a handful without red chips by choosing a handful of 40 from the 75 non-red chips. This is done in $\binom{75}{40}$ ways. Thus, $n(\text{at least 1 red chip}) = n(\text{no red chips}) =$

$$\binom{75}{40}.$$

2. Let A be the set of handfuls with at least 2 red chips. We count the complement. The A' = the set of handfuls of 40 chips with exactly 1 red chip or no red chips.

To count $n(\text{exactly 1 red})$ first draw 1 red chip from 25 in 25 ways. Take 39 white, blue, and black chips from 75 to complete your choice of 40. This is done in $\binom{75}{39}$ ways.

Choose a handful without red chips by choosing a handful of 40 from the 75 non-red chips. This is done in $\binom{75}{40}$ ways. Thus, $n(\text{at least 1 red}) = n(\text{exactly 1 red}) + n(\text{no red}) =$

$$25 \cdot \binom{75}{39} + \binom{75}{40}.$$

EXAMPLE 10.2.3 Choose a handful of 40 chips from the bag. Count the number of handfuls that contain at least 1 red and 1 white chip.

Solution: The set $\mathcal{U}$ is the set of all handfuls of 40 chips taken from 100 chips. Then $n(\mathcal{U}) = \binom{100}{40}$.

Let A be the set of handfuls of 40 chips that contain at least 1 red chip and at least 1 white chip. We examine the complement of A. By DeMorgan's law 2.4,

not(at least 1 red chip and at least 1 white chip)
= not(at least 1 red chip) or not(at least 1 white chip)
= no red chips or no white chips

and by the inclusion/exclusion principal we have

n(no red chips or no white chips) = n(no red chips)
$+n$(no white chips) $-$ n(no red chips and no white chips)

Then $n(A')$ = n(no red chips or no white chips) = n(no red chips) + n(no white chips) $-$ n(no red chips and no white chips)

The handfuls without red chips are found by taking a handful of 40 chips from the 75 white, blue, and black chips. This is done in $\binom{75}{40}$ ways.

The handfuls without white chips are found in a similar manner. Take a handful of 40 chips from the 75 red, blue, and black chips. This is also done in $\binom{75}{40}$ ways.

The handfuls that are without red chips and without white chips are found by taking a handful of 40 chips from the 50 blue and black chips. This is done in $\binom{50}{40}$ ways.

Then $n(A')$ = n(no red chips) + n(no white chips) $-$ n(no red chips and no white chips) =

$$2 \cdot \binom{75}{40} - \binom{50}{40}.$$

Therefore

$$n(A) = \binom{100}{40} - \left(2 \cdot \binom{75}{40} - \binom{50}{40}\right).$$

EXAMPLE 10.2.4 Choose a handful of 40 chips from the bag.

1. Count the number of possible handfuls with at most 24 red chips.

2. Count the number of possible handfuls with at most 23 red chips.

1. Observe that *at most 24 red chips* means *no red chips* or *exactly 1 red chip* or ... or *exactly 24 red chips*, which is a bit more work than we want to do right now. So use the complement formula.

Let $A =$ the set of handfuls of 40 chips with at most 24 red chips. Then $A' =$ the set of handfuls of 40 chips with at least 25 red chips. Because there are only 25 red chips, A' is the set of handfuls of 40 chips with exactly 25 red chips. The handfuls in A' are formed by choosing 25 red chips in exactly 1 way, and then choosing 15 non-red chips to fill out the 40 chips in the handful. We choose 15 non-red chips from the remaining 75 chips in $\binom{75}{15}$ ways. Thus, $n(A') = 1 \cdot \binom{75}{15}$. The set $\mathcal{U}$ of all 40 chip handfuls from the bag of 100 chips has exactly $\binom{100}{40}$ ways. Hence $n(A) =$

$$\binom{100}{40} - \binom{75}{15}.$$

2. This is done the same way that we did part 1. Let $A =$ be the set of handfuls of 40 chips that contain *at most 23 red chips.* Then $A' =$ the set of handfuls of 40 chips that contain *at least 24 red chips.* Since there are only 25 chips, $A' =$ the set of handfuls of 40 chips that contain *exactly 24 red chips* or *exactly 25 red chips.* As in part 1, the number of handfuls that contain exactly 25 red chips is $\binom{75}{15}$, while the number of handfuls with exactly 24 red chips is found by first choosing the 24 chips and then choosing the remaining 16 non-red chips. This can be done in

$$\binom{25}{24}\binom{75}{16} = 25 \cdot \binom{75}{16}$$

ways. Thus,

$$n(A') = \binom{75}{15} + 25 \cdot \binom{75}{16}.$$

If we let $\mathcal{U}$ = the set of handfuls of 40 chips, then $n(\mathcal{U}) = \binom{100}{40}$, and therefore

$$n(A) = \binom{100}{40} - \binom{75}{15} - 25 \cdot \binom{75}{16}.$$

EXAMPLE 10.2.5 Choose a handful of 40 chips. Count the number of possible handfuls with at most 24 chips of one color.

There are 4 ways to choose the color, say red. The handful must have at most 24 chips of that color. Then the problem is to count the number of handfuls with at most 24 red chips. This is Example 10.2.4, where we found that the number of such handfuls is

$$\binom{100}{40} - \binom{75}{15}.$$

Then the total number of handfuls of 40 chips containing at most 24 chips of some color is

$$4 \cdot \left(\binom{100}{40} - \binom{75}{15}\right).$$

10.3 Exercises

Five is small. Do these problems in two ways. These cards are from the standard deck of 52 cards.

1. Count the number of 5 card hands that have at least 1 heart.

Answer: $\binom{52}{5} - \binom{39}{5}$

2. Count the number of 5 card hands that have at most 4 hearts.

Answer: $\binom{52}{5} - \binom{13}{5}$

3. Count the number of 5 card hands that have at most 3 hearts.

Answer: $\binom{52}{5} - \binom{13}{4} - \binom{13}{5}$

4. Count the number of 5 card hands that have at least 1 heart or at least 1 diamond. Answer: $\binom{52}{5} - \binom{26}{5}$

5. Count the number of 5 card hands that have at least 1 Ace and at least 1 King. Answer: $\binom{52}{5} - 2 \cdot \binom{48}{5} + \binom{44}{5}$

Alternative answer:

$$4^2 \cdot \binom{44}{3} + 4 \cdot \binom{4}{2}\binom{44}{2} + \binom{4}{2}^2 \cdot 44 + 4^2 \cdot 44 + 4$$

6. Count the number of 5 card hands that have at least 1 heart and at least 1 diamond. Answer: $\binom{52}{5} - 2 \cdot \binom{39}{5} + \binom{26}{5}$

7. Count the number of 3 card hands that have three different kinds of cards. Answer: $\binom{13}{3} \cdot 4^3$

Alternative answer: $\binom{52}{3} - 13 \cdot \binom{4}{2} \cdot 48 - 13 \cdot 4$

Chapter 11

Advanced Permutations

In this chapter we will combine several methods from earlier chapters and use them to solve more advanced counting problems. We will use permutation problems and Venn diagrams to solve some counting problems that otherwise would be too hard to call elementary combinatorics. These problems will force the reader to divide a larger problem into smaller, more manageable problems. In this way, the reader will be doing mathematics the way the professionals do it.

11.1 Venn Diagrams and Permutations

In the first two problems we examine words that contain exactly 1 a and exactly 1 b. The use of a and b here is generic. We could have used any two different symbols in any finite alphabet of symbols. The techniques would be the same. The use of 3-letter words is meant to simplify the counting process. We could just as easily used many more letters. The problems would be the same whether we use 3 or 4 letters, but they get harder as we use more letters. The best way to explain this is to do a few examples and to leave similar problems to the exercises.

EXAMPLE 11.1.1 We choose a word from the set of 3-letter words from the English alphabet $\{a, \dots, z\}$. Observe the occurrence of a and b in the word. Fill in the appropriate 2-circle Venn diagram if we observe exactly 1 a or exactly 1 b in the word.

Solution: Let $\mathcal{U}$ be the set of all 3-letter words. In the Venn diagram let A be the set of 3-letter words in which a occurs exactly once. Let B be the set of 3-letter words in which b occurs exactly once. We count each simple component. We will need to know that the universal set contains $n(\mathcal{U}) = 26^3$ words. We will identify the number of elements in a region by one of the letters a, b, c, d as in the Venn diagram.

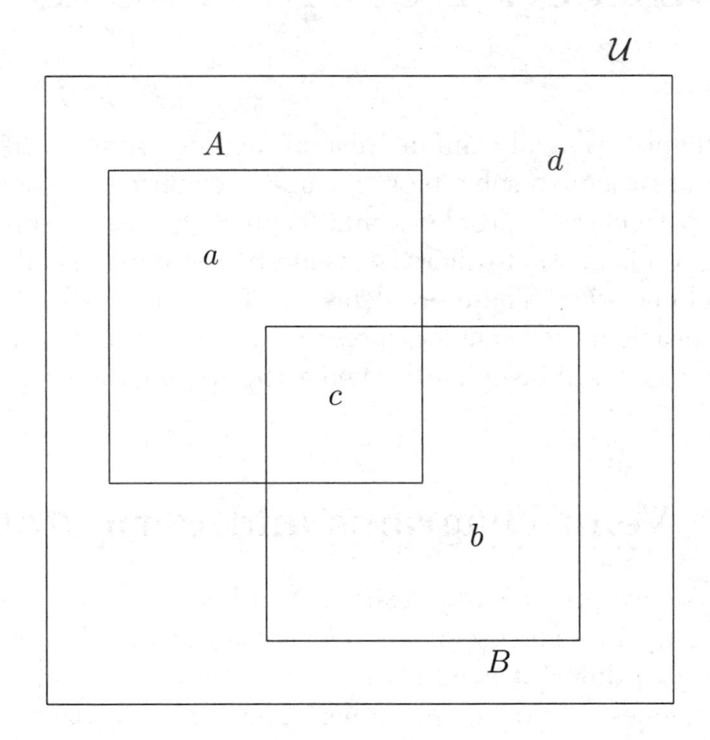

A word in $A \cap B'$ contains exactly 1 a and no b's. We choose 1 place in 3 for the a, which is done in 3 ways. Then fill in each of the remaining 2 places with one of 24 letters not equal a or b. This is done in 24 ways. Then $a = n(A \cap B') = 3 \cdot 24^2$.

The number $n(B \cap A')$ is calculated in the same way, so $b = n(B \cap A') = 3 \cdot 24^2$.

The set $c = A \cap B$ is all 3-letter words with exactly 1 a and exactly 1 b. Choose a place for the a in 3 ways, choose a place for the b in 2 ways, and fill the remaining place with 1 of 24 letters $\{c, d, \ldots, z\}$ in 24 ways. Then $n(A \cap B) = 3 \cdot 2 \cdot 24$.

By DeMorgan's Law, $(A \cup B)' = A' \cap B'$ is the set of 3-letter words that do not contain an a and do not contain a b. Then $(A \cup B)'$ is the set of 3-letter words from the alphabet $\{c, d, \ldots, z\}$ consisting of 24 letters. Hence $d = n(A \cup B)' = 24^3$.

The Venn diagram is then complete.

Notice that $A \cup B$ is the set of 3-letter words containing exactly 1 a or exactly 1 b. We find $n(A \cup B)$. By the Venn diagram we see that

$$A \cup B = (A \cup B') \cup (A' \cup B) \cup (A \cap B)$$

and these regions are disjoint. Thus

$$\begin{aligned} n(A \cup B) &= n(A \cup B') + n(A' \cup B) + n(A \cap B) \\ &= 2(3 \cdot 24^2) + 3 \cdot 2 \cdot 24. \end{aligned}$$

The reader might try finding $n(A \cup B)$ using a more elementary approach.

So problems in which we are counting words with *exactly 1* letter, are solved with the above procedure. You do a simple permutation problem several times to fill in the connected regions of the Venn diagram.

Here is a problem that uses words that contain *at least 1 a*. As above, you will do a simple permutation problem several times to fill in the Venn diagram.

EXAMPLE 11.1.2 Choose a word from the set of 3-letter words from the English alphabet $\{a, \ldots, z\}$. Observe the occurrence of a and b in the word. Fill in the appropriate Venn diagram if we observe that the word contains at least 1 a or at least 1 b.

Solution: The letters A, B, and a, b, c, d refer to the Venn diagram on page 145. Let A denote the set of 3-letter words with at least 1 a. Let B denote the set of 3-letter words with at least 1 b.

The usual method of counting shows us that

$$n(\mathcal{U}) = 26^3.$$

$A \cap B'$ is the set of all 3-letter words containing at least 1 a and no b's. Our experience shows us that it is better to count the complement of $A \cap B'$. Observe that $A \cap B'$ is contained in the set of 3-letter words from the 25 letter alphabet $\{a, c, d \ldots, z\}$. This is the English alphabet less the letter b. Let $\mathcal{V}$ be the 3-letter words from the 25 letter alphabet $\{a, c, d \ldots, z\}$. Then $n(\mathcal{V}) = 25^3$.

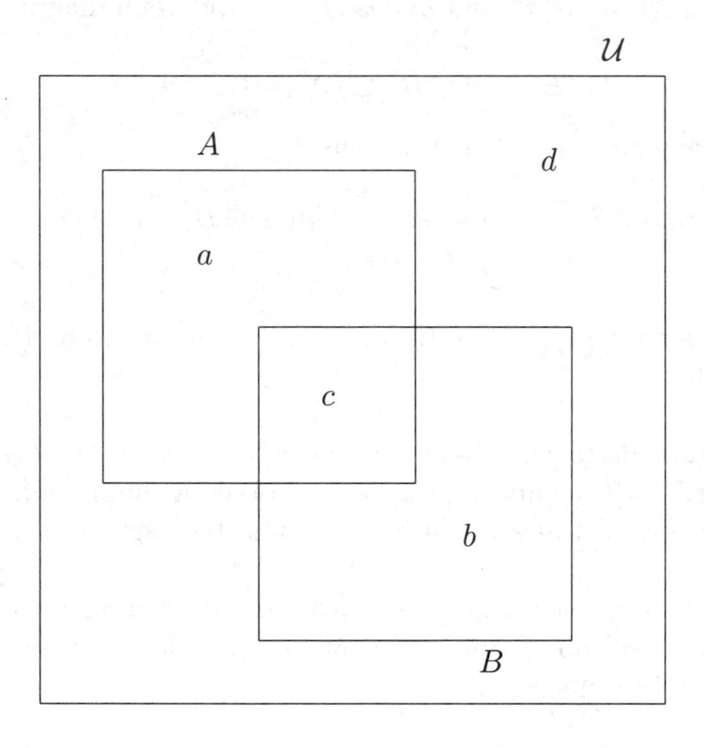

To count the words in $\mathcal{V}$ that do not contain an a, count the 3-letter words from the alphabet $\{c, d, \ldots, z\}$. These are the words that do not contain a or b. There are 24^3 such letters. Similarly calculate b. Then

$$a = n(A \cap B') = 25^3 - 24^3 \quad \text{and} \quad b = n(B \cap A') = 25^3 - 24^3.$$

Now take an unexpected path. Count $(A \cup B)'$. The set $A \cup B$ consists of the words containing at least 1 a or 1 b. By DeMorgan's Law, $(A \cup B)' = A' \cap B'$ is the set of 3-letter words that contain neither a nor b. Then $(A \cup B)'$ is the set of 3-letter words from the 24 letter alphabet $\{c, d, \ldots, z\}$. As usual, there are 24^3 such words, so that

$$d = n(A \cup B)' = 24^3.$$

Find $c = n(A \cap B)$. Adding up the regions in the diagram we get

$$n(\mathcal{U}) = 2 \cdot n(A \cup B') + n(A \cap B) + n(A \cup B)'$$

so that

$$26^3 = 2(25^3 - 24^3) + n(A \cap B) + 24^3.$$

Hence,

$$c = n(A \cap B) = 26^3 - 2 \cdot 25^3 + 24^3.$$

This completes the diagram.

As an application, let us find $n(A \cup B)$. By adding up the regions in the diagram we have

$$n(A \cup B) = 2(25^3 - 24^3) + 26^3 - 2 \cdot 25^3 + 24^3 = 26^3 - 24^3.$$

Or we could have used the complement formula noting from the diagram that

$$n(A \cup B) = n(\mathcal{U}) - n(A \cup B)' = 26^3 - 24^3.$$

You may ask why we took the unexpected path in the above problem. The next problem explains why. We find the solution in the above problem has an elegance about it that is not possessed by a solution involving the more direct solution of taking many cases. To illustrate the point, we solve the problem again using that more direct solution.

EXAMPLE 11.1.3 Choose a 3-letter word from the alphabet $\{a, \ldots, z\}$. Observe the occurrence of a and b in the word. Fill in the appropriate Venn diagram if we observe at least 1 a or at least 1 b in the word.

Solution: Let $\mathcal{U}$ be the set of 3-letter words from $\{a, \ldots, z\}$. Let A denote the set of 3-letter words with at least 1 a. Let B denote the set of 3-letter words with at least 1 b.

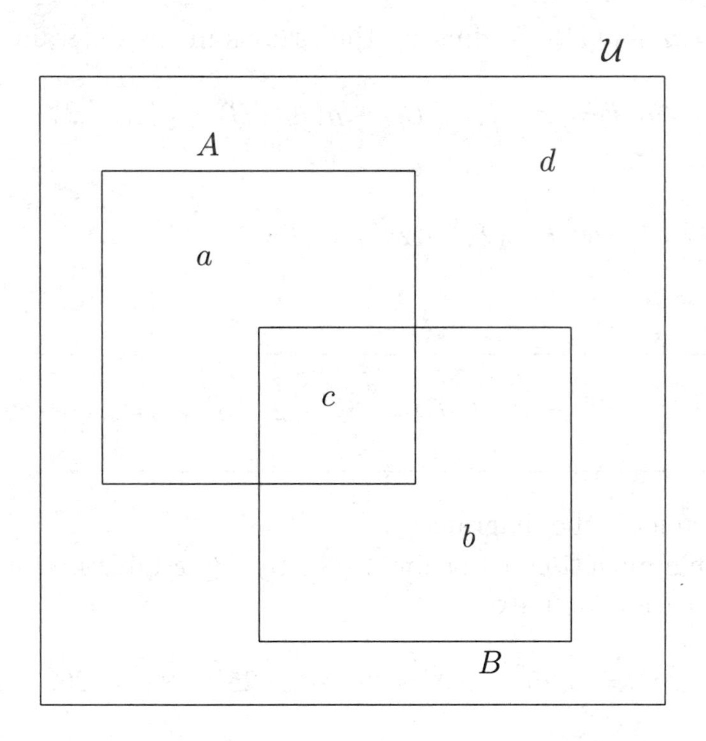

As usual

$$n(\mathcal{U}) = 26^3.$$

We proceed as we did in earlier Venn diagrams, counting the smallest region in the diagram. $A \cap B$ is the set of all 3-letter words with at least 1 a and at least 1 b. There are several cases to cover.

1. The words in this case contains exactly 1 a and exactly 1 b. Choose a place for an a in 3 ways, and choose a place for a b in

2 ways. Choose a letter from $\{c, d, \ldots, z\}$ in 24 ways for the third place. That is, choose a letter that is not an a or a b. Combining these we see that there are $3 \cdot 2 \cdot 24$ such words.

2. The words in this case contain exactly 2 a's and exactly 1 b. This accounts for all of the 3-letters in the word. Choose 1 of 3 places for the b and the rest is for a. This is done in 3 ways. Similarly, there are 3 words that contain exactly 2 b's and 1 a. Combining, we find that this case contributes $3 + 3 = 6$ words to $A \cap B$.

3. The words with exactly 3 a's or b's are aaa and bbb. Thus no words in this case contain at least 1 a and at least 1 b. These words do not contribute to our solution.

Adding up all of the cases we see that

$$c = n(A \cap B) = 3 \cdot 2 \cdot 24 + 6 = 6 \cdot 25.$$

From the previous example,

$$a = n(A \cap B') = b = n(B \cap A') = 25^3 - 24^3.$$

Hence

1. $n(A \cup B) = 2(25^3 - 24^3) + 6 \cdot 25.$
2. $d = n(A \cup B)' = 26^3 - 2(25^3 - 24^3) - 6 \cdot 25$ so that

$$d = 26^3 - 2 \cdot 25^3 + 2 \cdot 24^3 - 6 \cdot 25.$$

Watch what happens to the number of cases needed to count $c = n(A \cap B)$ when we increase the number of letters in the words from 3 to 4.

EXAMPLE 11.1.4 Choose a word from a set of 4-letter words from the alphabet $\{a, \ldots, z\}$. Observe the occurrence of a and b in the word. Fill in the appropriate 2 circle Venn diagram if we observe at least 1 a or at least 1 b in the word.

Solution: The number of 4-letter words from a 26 letter alphabet is

$$n(\mathcal{U}) = 26^4.$$

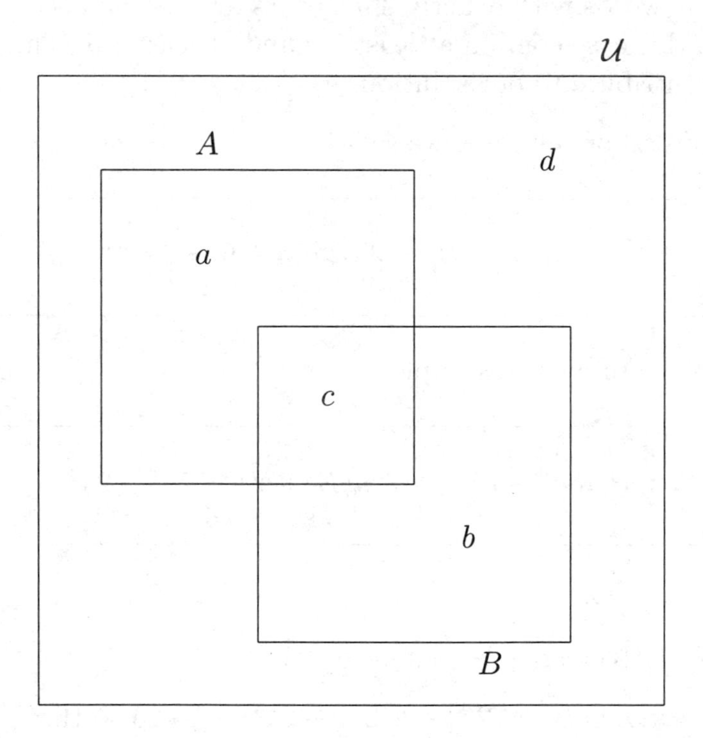

Let A denote the set of 4 letter words with at least 1 a. Let B denote the set of 4 letter words with at least 1 b. We employ the complement formula.

$A \cap B'$ is the set of all 4 letter words containing at least 1 a and no b's. These are words from the set of 4 letter words from the 25 letter alphabet $\{a, c, d, \ldots, z\}$. (Note the missing b.) There are 25^4 four letter words from $\{a, c, d, \ldots, z\}$. The words in the complement

$(A \cap B')'$ are 4 letter words not containing a (or b). These are 4 letter words from the 24 letter alphabet $\{c, d, \ldots, z\}$, and there are 24^4 such words. The same number of elements is in $B \cap A'$. Thus

$$a = n(A \cap B') = 25^4 - 24^4 = b = n(B \cap A').$$

$(A \cup B)$ is the set of 4 letter words with at least 1 a or at least 1 b, so by DeMorgan's law, $(A \cup B)' = A' \cap B'$ is the set of 4 letter words without a's and without b's. These are the 4 letter words from the alphabet $\{c, d, \ldots, z\}$, and there are

$$d = n(A \cup B)' = 24^4$$

such words. Hence

$$n(A \cup B) = n(\mathcal{U}) - n(A \cup B)' = 26^4 - 24^4.$$

Hence, by looking at the Venn diagram, $c = n(A \cap B) = n(A \cup B) - n(A \cap B') - n(B \cap A') = (26^4 - 24^4) - 2(25^4 - 24^4)$, so that

$$c = n(A \cap B) = 26^4 - 2 \cdot 25^4 + 24^4.$$

This completely labels the Venn diagram.

In the above Example, let us see how to find $n(A \cap B)$ using cases. Since $A \cap B$ is the set of 4 letter words with at least 1 a and at least 1 b, the possible cases for the number of a's and b's in $A \cap B$ are

exactly 1 a and exactly 1 b
exactly 2 a's and exactly 1 b
exactly 3 a's and exactly 1 b
exactly 2 a's and exactly 2 b's.

Moreover, we leave it to the reader to show that

$$n(\text{exactly 2 } a\text{'s and exactly 1 } b) = n(\text{exactly 1 } a\text{'s and exactly 2 } b)$$
$$n(\text{exactly 3 } a\text{'s and exactly 1 } b) = n(\text{exactly 1 } a\text{'s and exactly 3 } b)$$

These two lists then account for every case in $A \cap B$ for the number of a's and number of b's in a word. Thus the number of words $c = n(A \cap B)$ is given by the sum

$$\begin{aligned} c = n(A \cap B) &= n(\text{exactly } 1\ a \text{ and exactly } 1\ b) \\ &+ 2n(\text{exactly } 2\ a\text{'s and exactly } 1\ b) \\ &+ 2n(\text{exactly } 3\ a\text{'s and exactly } 1\ b) \\ &+ n(\text{exactly } 2\ a\text{'s and exactly } 2\ b\text{'s}). \end{aligned}$$

Then to find $c = n(A \cap B)$ we will analyze these four different cases using the methods we have exploited so far. Without details, your answer would look like this:

$$c = n(A \cap B) = 4 \cdot 3 \cdot 24^2 + 2 \cdot \binom{4}{2} \cdot 3 \cdot 24 + 2 \cdot 4^2 + \binom{4}{2}^2$$

The problems grow in complexity as we increase the number of occurrences of a and b in the words.

EXAMPLE 11.1.5 Choose a word from the set of 4 letter words from the alphabet $\{a, \ldots, z\}$. Observe the occurrence of a and b in the word. Fill in the appropriate 2 circle Venn diagram if we observe exactly 1 pair of a's or exactly 1 pair of b's in the word.

Solution: Let $\mathcal{U}$ be the set of all 4 letter words. Then

$$n(\mathcal{U}) = 26^4.$$

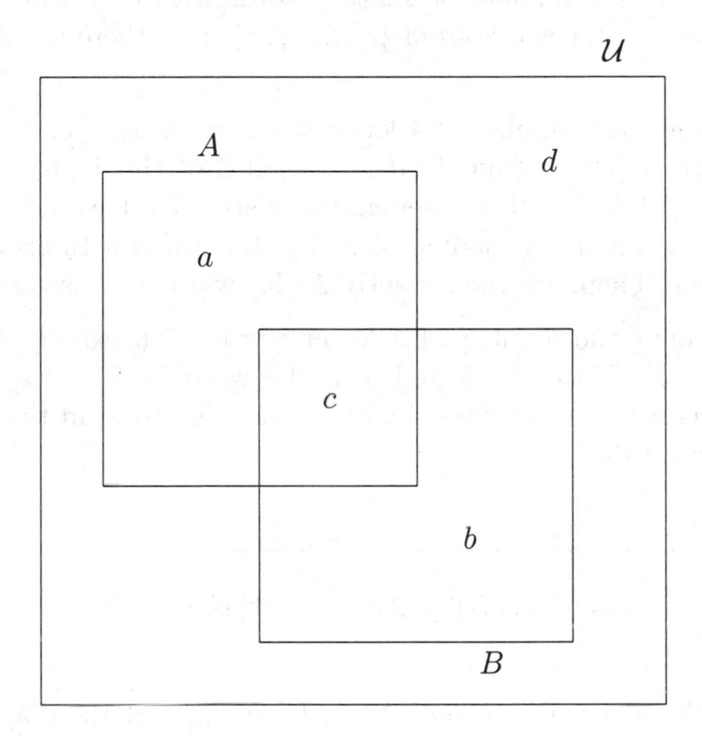

Let A be the set of words in which a occurs exactly 2 times. Let B be the set of words in which b occurs exactly 2 times.

$A \cap B$ is the set of 4 letter words with a pair of a's and a pair of b's. Since this word has exactly 4 letters, these are all of the letters in the word. There are $\binom{4}{2}$ ways to place the 2 a's and then 1 way to place the 2 b's. Thus

$$c = n(A \cap B) = \binom{4}{2}.$$

Count $(A \cup B)'$. By DeMorgan's law, $(A \cup B)' = A' \cap B'$ consists of 4 letter words without repeated a's and without repeated b's. There are several cases to consider. We employ the complement method.

1. Count the number of 4 letter words with no a's and no b's. These are 4 letter words from $\{c, d, \ldots, z\}$, and there are 24^4 such words.

2. Count the number of 4 letter words with no a's and exactly 1 b. Place 3-letters from $\{c, d, \ldots, z\}$ in 3 of the 4 places in 24^3 ways. The 1 b goes in the remaining place. We have constructed 24^3 words. Similarly, there are 24^3 4 letter words with exactly 1 a and no b's. There are then exactly $2 \cdot 24^3$ words in this case.

3. Count the number of 4 letter words with exactly 1 a and exactly 1 b. Place the a and b in the word in $4 \cdot 3$ ways. The remaining 2 places are filled in 24^2 ways. The total in this case is $4 \cdot 3 \cdot 24^2$ words.

Thus

$$d = n(A \cup B)' = 24^4 + 2 \cdot 24^3 + 4 \cdot 3 \cdot 24^2.$$

Finally, $n(A \cap B') = n(A' \cap B)$. By looking at the diagram we see that

$$n(\mathcal{U}) = n(A \cup B)' + 2n(A \cup B') + n(A \cap B)$$

so that

$$26^4 = 24^4 + 2 \cdot 24^3 + 4 \cdot 3 \cdot 24^2 + 2n(A \cap B') + \binom{4}{2}.$$

Solving for $n(A \cap B')$ we find the missing parts of the Venn diagram.

$$\begin{aligned} a = n(A \cap B') &= \frac{1}{2}\left(26^4 - 24^4 - 2 \cdot 24^3 - 4 \cdot 3 \cdot 24^2 - \binom{4}{2}\right) \\ &= \frac{1}{2}\left(26^4 - 24^4 - 2 \cdot 24^3 - 12 \cdot 24^2 - 6\right). \end{aligned}$$

The last region is calculated as $n(A \cap B') = n(A' \cap B) = b$.

In the last step of the above example you might have tried cases again. To find $n(A \cap B')$ you note that $A \cap B'$ is the set of 4 letter words that contain at least 2 a's, and that contain at most 1 b. The cases considered are as follows.

exactly 2 a's and no b's
exactly 3 a's and no b's
exactly 4 a's
exactly 2 a's and 1 b
exactly 3 a's and 1 b's

You would not consider 2 b's in a word since we are working in B', in which there is at most 1 b. Moreover, some of these cases count a different region in the Venn diagram. Observe.

n(exactly 2 a's and no b's) $=$ n(no a's and exactly 2 b's)
n(exactly 3 a's and no b's) $=$ n(no a's and exactly 3 b's)
n(exactly 4 a's) $=$ n(exactly 4 b's)
n(exactly 2 a's and 1 b) $=$ n(exactly 1 a and exactly 2 b's)
n(exactly 3 a's and 1 b) $=$ n(exactly 1 a and exactly 3 b's)

This is probably more counting than we want to do to find $n(A\cap B')$. Thus, when counting $A\cap B'$ in problems like the one above, I suggest that we use the diagram as we did at the end of the above example.

By replacing *at least* with *at most* we have found another problem that requires yet more cases to solve.

EXAMPLE 11.1.6 Choose 1 word from the set of 4 letter words from the English alphabet $\{a,\ldots,z\}$. Observe the occurrence of a and b in the word. Fill in the appropriate 2 circle Venn diagram if we observe at most 1 a or at most 1 b in the word.

Solution: Let A denote the set of 4 letter words with at most 1 a, and let B denote the set of all 4 letter words with at most 1 b. The possible occurrences of a or b are *not at all* or *exactly once.*

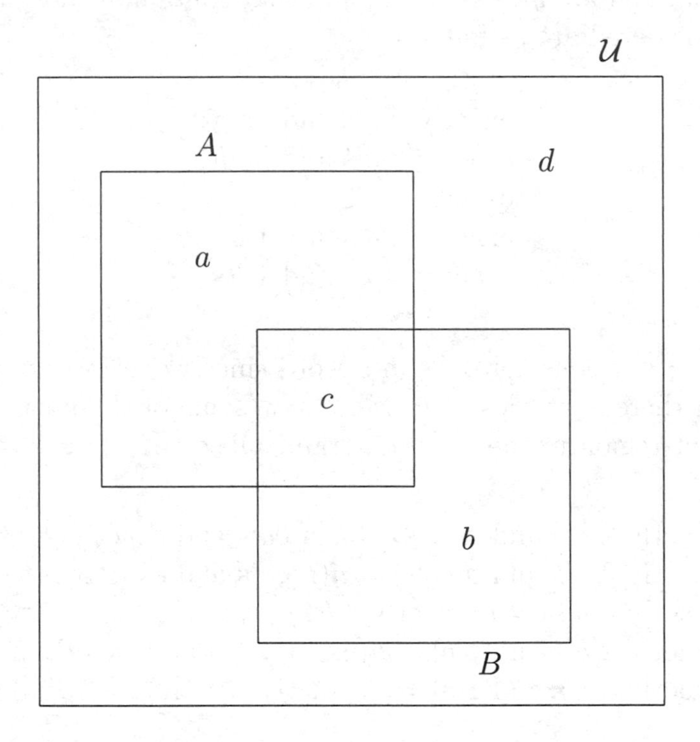

As usual

$$(\mathcal{U}) = 26^4.$$

Count $d = (A \cup B)'$. By DeMorgan's law $(A \cup B)' = A' \cap B'$ is the set of words with at least 2 a's and at least 2 b's in a 4 letter word. This word has exactly 2 a's and 2 b's, and there are exactly

$$d = n(A \cup B)' = \binom{4}{2} = 6$$

such words. (Choose 2 places in 4 for the a's.)

$A \cap B$ is the set of words with at most 1 a and at most 1 b. The three cases here are

no a's and no b's in a word,
exactly 1 a and no b's in a word, and
exactly 1 a and 1 b in a word.

and we observe that

$$n(\text{exactly } 1\ a \text{ and no } b\text{'s in a word}) = n(\text{no } a\text{'s and exactly } 1\ b \text{ in a word}).$$

Now $n(\text{no } a\text{'s and no } b\text{'s in a word}) = 24^4$ since a word with no a's or b's is a word from the alphabet $\{c, d, \ldots, z\}$. The words with exactly 1 a and no b's are words from the 25 letter alphabet $\{a, c, d, \ldots, z\}$ that contain exactly 1 a. When we count, we find that

$$n(\text{exactly } 1\ a \text{ and no } b\text{'s in a word}) = 4 \cdot 24^3.$$

The words with exactly 1 a and 1 b have 2 places to fill with 24 letters. Thus

$$n(\text{exactly } 1\ a \text{ and } 1\ b \text{ in a word}) = 4 \cdot 3 \cdot 24^2.$$

Adding up these cases we find that

$$\boxed{c = n(A \cap B) = 24^4 + 2 \cdot 4 \cdot 24^3 + 4 \cdot 3 \cdot 24^2.}$$

Observe that $a = n(A \cap B') = n(A' \cap B) = b$. By adding the regions in the Venn diagram we see that

$$n(\mathcal{U}) = n(A \cup B)' + 2n(A \cap B') + n(A \cap B).$$

Substitute the values we found above into this sum and find that

$$\begin{aligned} 26^4 &= 6 + 2n(A \cap B') + 24^4 + 8 \cdot 24^3 + 12 \cdot 24^2 \\ n(A \cap B') &= \frac{1}{2}\left(26^4 - (6 + 24^4 + 8 \cdot 24^3 + 12 \cdot 24^2)\right) \end{aligned}$$

and so

$$a = b = \frac{1}{2}\left(26^4 - 24^4 - 8 \cdot 24^3 - 12 \cdot 24^2 - 6\right).$$

This completes the Venn diagram.

Problems that correspond to 3-square Venn diagrams are handled in the same manner. We leave them as exercises.

11.2 Exercises

Problems 1-4 use the Venn diagram (12.1.4) in section 1.

In exercises 1 through 4 choose a word from the set of 4-letter words, and observe the occurrence of a and b in the word.

1. Fill in the appropriate Venn diagram if we observe exactly 2 a's or exactly 2 b's in the word. Answer: $c = \binom{4}{2}$, $a = b = \binom{4}{2}(24^2 + 2 \cdot 24)$, $d = 26^4 - \binom{4}{2}(24^2 + 2 \cdot 24 + 1)$.

2. Fill in the appropriate Venn diagram if we observe at least 1 a or at least 1 b.

3. Fill in the appropriate Venn diagram to represent the fact that you will choose a word containing at least 2 a's or at least 2 b's. Answer: $c = \binom{4}{2}$, $d = 24^4 + 2(4 \cdot 24^3 + 4 \cdot 3 \cdot 24^2)$, $a = b = \frac{1}{2}(26^4 - c - d)$.

4. Fill in the appropriate 2-circle Venn diagram if we observe exactly 2 of one letter or exactly 2 of another letter.

The Venn diagrams used in these exercises have three squares. The letters a, b, c, d, e, f, g, h refer to the number of elements in the corresponding regions in the above Venn diagram. The words are from portions of the English alphabet.

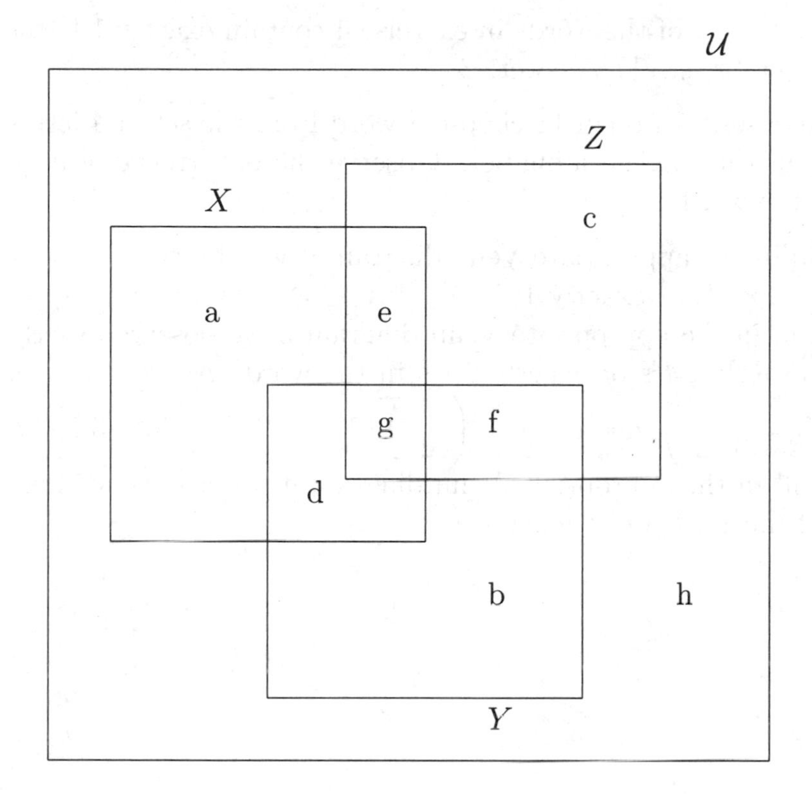

5. You have a set of 3 letter words from the alphabet $\{u, v, w, x, y, z\}$. Fill in the appropriate Venn Diagram to represent the fact that you will choose a word with at least 2 x's or at least 2 y's or at least 2 z's. Answer: $d = e = f = g = 0$, $a = b = c = 3 \cdot 5 + 1$, $h = 6^3 - 45$.

6. In exercise 5, how many words do not contain at least 2 x's, or at least 2 y's, or at least 2 z's? Answer: $6^3 - 3 \cdot (3 \cdot 5 + 1)$.

In exercises 7 and 8 choose a word from the set of 3-letter words from the alphabet $\{u, v, w, x, y, z\}$. Observe the occurrence of a, b, and c in the word.

7. Fill in the appropriate Venn Diagram to represent the fact that you will choose a word with exactly 1 x or exactly 1 y or exactly 1 z. Answer: $h = 3^3$, $g = 1$, $d = e = f = 4 \cdot 3 \cdot 2$, $a = b = c = \frac{1}{3}\left(6^3 - 3^3 - 3 \cdot 24 - 1\right)$.

8. How many of the words in exercise 4 contain exactly 1 letter from the set $\{x, y, z\}$? Answer: 3^3

In exercises 9 through 11 choose a word from the set of 4 letter words from the English alphabet. Observe the occurrence of a, b, and c in the word.

9. Fill in the appropriate Venn diagram if we observe exactly 1 a, or exactly 1 b, or exactly 1 c.

10. Fill in the appropriate Venn diagram if we observe exactly 2 a's or exactly 2 b's or exactly 2 c's in the word. Answer: $g = 0$, $d = e = f = \binom{4}{2}$, $a = b = c = \binom{4}{2}(23^2 + 2)$, $h = 26^4 - 3a - 3d$.

11. Fill in the appropriate Venn diagram if we observe at most 1 a, or at most 1 b, or at most 1 c.

Chapter 12

Advanced Combinations

We will use combination problems and Venn diagrams to solve some counting problems that otherwise would be too hard to call elementary combinatorics. Some of the problems we encounter will be unusual combination problems. In this way, we increase the educational demands on the reader. However, these problems are still within the grasp of the average college freshmen.

12.1 Venn Diagrams and Combinations

The type of problem that we will study here begins with two properties P and Q that a set might possess, and then count the number ways that a set can possess the properties P and Q. We may even improve the problem to include a third property. We begin with a simple setting.

NOTATION 12.1.1 We are given a bag full of colored numbered chips. There are 13 red round chips numbered 1, 2, 3, 4, 5, 6, 7, 8, 9, 10, 11, 12, 13. There are 13 red square chips numbered from 1 to 13. There are 13 black round chips numbered 1 to 13. There are 13 black square chips numbered 1 to 13. Any mention of a bag in the following examples refers to this bag.

EXAMPLE 12.1.2 We choose 3 chips from the bag. Label a Venn diagram to reflect the fact that our handful of chips contains exactly 1 red square chip or exactly 1 black square chip.

Solution: Let $\mathcal{U}$ be the set of handfuls of 3 chips from the bag. Let A be the set of handfuls that contain exactly 1 red square chip. Let B be the set of handfuls that contain exactly 1 black square chip.

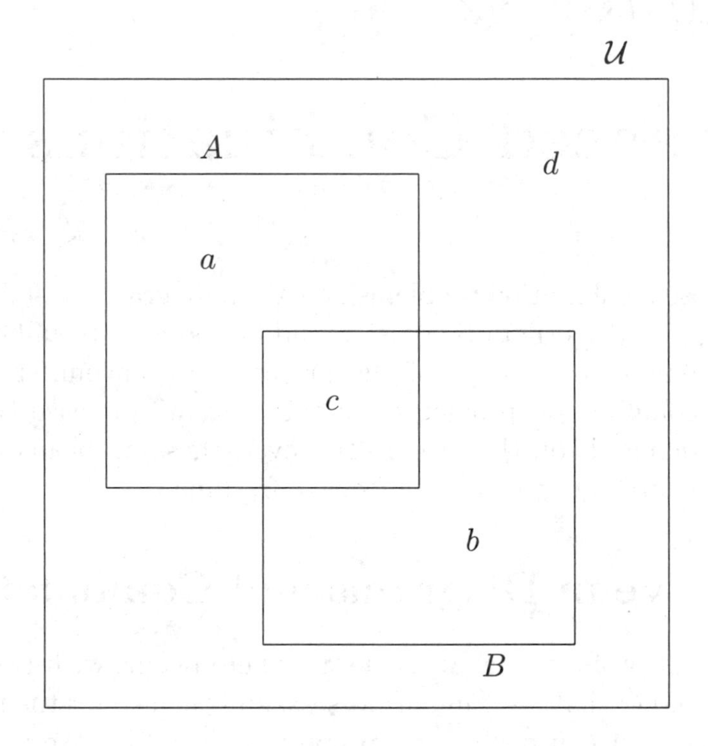

To count $\mathcal{U}$ we are choosing 3 chips from a bag of 52 different chips. This is the prototypical combination problem. Then $n(\mathcal{U}) = \binom{52}{3}$ handfuls.

We begin by labeling the smallest region.

$A \cap B$ is the set of 3 chip handfuls that have exactly 1 red square chip and 1 black square chip. Form a handful as follows. Choose 1 red square chip, and then choose 1 black square chip. Each choice is made in 13 ways. The remaining chip is found in the 26 round chips, which can be chosen in 26 ways. Then

$$n(A \cap B) = 13 \cdot 13 \cdot 26.$$

To count $A \cap B'$, break the handfuls into two cases. Each handful must have exactly 1 red square chip since the set $A \cap B'$ is in A.

1. The first case is a handful with exactly 1 red square chip and no black square chips. This is formed by choosing 1 of 13 red square chips and then choosing 2 chips from the 26 round chips. This is done in $13 \cdot \binom{26}{2}$ ways.

2. Since $A \cap B'$ is in A and B', we choose a handful with exactly 1 red square chip and exactly 2 black square chips. (Since the handful is in B', we do not have exactly 1 black square chip.) Since the handful consists of exactly 3 chips, this is the only other case. Choose 1 red square chip in 13 ways, and choose 2 of 13 black square chips in $\binom{13}{2}$ ways. Then this case contains exactly $13 \cdot \binom{13}{2}$ handfuls.

Then

$$\begin{aligned} n(A \cap B') &= 13 \cdot \binom{26}{2} + 13 \cdot \binom{13}{2} \\ &= 13^2 \cdot 31 \end{aligned}$$

By using the Venn diagram, we see that

$$\begin{aligned} n(A \cup B) &= n(A \cap B') + n(A \cap B) + n(A' \cap B) \\ &= 2n(A \cap B') + n(A \cap B) \\ &= 2(13^2 \cdot 31) + 13^2 \cdot 26 \\ &= 13^2 \cdot 88 \end{aligned}$$

The reader is encouraged to figure out where these numbers came from.

Since $n(\mathcal{U}) = n(A \cup B)' + n(A \cup B)$ we have

$$n(A \cup B)' = \binom{52}{3} - 13^2 \cdot 88$$

handfuls. This completes the Venn diagram.

Notice that in this problem we were doing the same combination problem over and over with different numbers. You would expect this as we are finding all possible handfuls containing various combinations of chips. Here is another.

EXAMPLE 12.1.3 Choose a handful of 4 chips from the bag. Fill in the appropriate Venn diagram

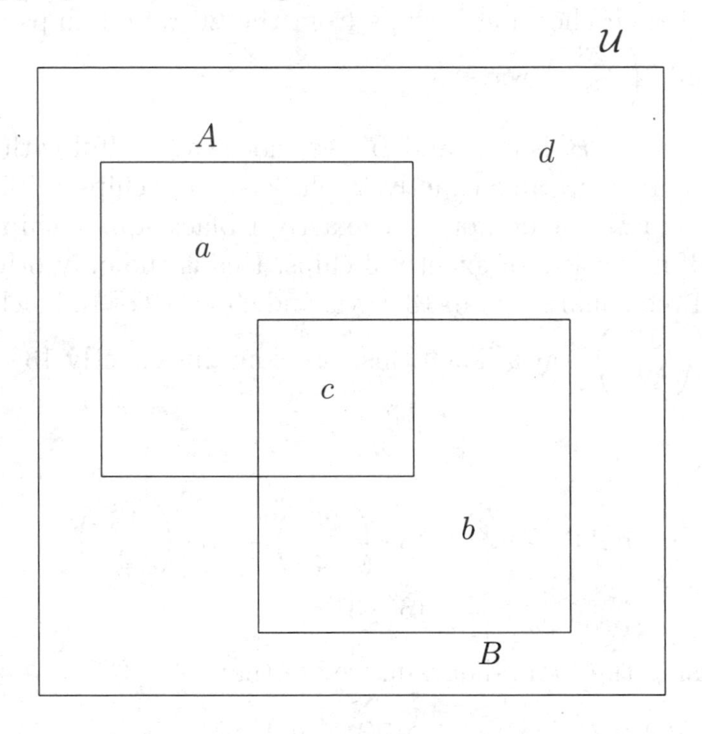

corresponding to the fact that the chosen handful contains at least 1 red square chip or at least 1 black square chip.

Solution: Let $\mathcal{U}$ be the set of handfuls of 4 chips from our bag. Let A be the set of handfuls that contain exactly 1 red square chip. Let B be the set of handfuls that contain exactly 1 black square chip.

To form $\mathcal{U}$ we are choosing 4 chips from the bag of 52 chips. This is a combination problem, so

$$n(\mathcal{U}) = \binom{52}{4}.$$

We break the counting into three cases. 1. By DeMorgan's law, $(A \cup B)' = A' \cap B'$ is the set of handfuls of 4 chips that does not contain a red square chip and does not contain a black square chip. These are handfuls of 4 chips from the 26 round chips. Thus,

$$n(A \cup B)' = \binom{26}{4}.$$

Then by the complement formula we find that

$$\begin{aligned} n(A \cup B) &= n(\mathcal{U}) - n(A \cup B)' \\ &= \binom{52}{4} - \binom{26}{4}. \end{aligned}$$

2. $A \cap B'$ is the set of handfuls of 4 chips that contain at least 1 red square chip and no black square chips. These handfuls come from a different alphabet, which we call X. The black square chips are missing from X, and X contains the other chips from the bag. Thus, there are 39 chips in X, so that

$$\binom{39}{4}$$

is the number of handfuls of 4 chips from X.

Observe that $A \cap B'$ is a set of handfuls from X. The complement of $A \cap B'$ consists of those handfuls of 4 chips from X without a red square chip. Since X does not contain black square chips, the complement of $A \cap B'$ is the set of handfuls of 4 chips from the set of 26 round chips. Thus, there are

$$\binom{26}{4}$$

such handfuls. Then by the complement formula

$$n(A \cap B') = \binom{39}{4} - \binom{26}{4}.$$

Then

$$\boxed{a = b = n(B \cap A') = \binom{39}{4} - \binom{26}{4}.}$$

3. Since $n(A \cap B') = n(B \cap A')$ and by reading the Venn diagram, we find that

$$\begin{aligned} n(A \cap B) &= n(A \cup B) - 2n(A \cap B') \\ &= \binom{52}{4} - \binom{26}{4} - 2\left(\binom{39}{4} - \binom{26}{4}\right) \end{aligned}$$

and so

$$c = \binom{52}{4} - 2\binom{39}{4} + \binom{26}{4}.$$

This completes the Venn diagram.

One advantage to these problems is that we have done several problems at once. In this way we justify the work involved in making such a Venn diagram.

EXAMPLE 12.1.4 Choose a handful of 5-chips from the bag. Fill in the appropriate Venn diagram corresponding to the observation that the chosen handful contains at least 2 chips numbered with 12 or at least 2 chips numbered with 13.

Solution: The set A consists of 5-chip handfuls from the bag that contain at least 2 chips numbered 12. The set B consists of 5-chip handfuls from the bag that contain at least 2 chips numbered 13. To fill in the Venn diagram requires us to evaluate several cases.

The set $\mathcal{U}$ is all possible handfuls of 5-chips from our bag, so

$$n(\mathcal{U}) = \binom{52}{5}.$$

We observe that $(A \cup B)'$ is the set of all 5-chip handfuls that have at most 1 chip numbered 12 and at most 1 chip numbered 13. The cases follow.

1. A 5-chip handful with no chips numbered 12 or 13 is formed by choosing a 5-chip handful from the bag ignoring the 8 chips numbered 12 or 13. There are 44 such chips other than those numbered 12 or 13. There are $\binom{44}{5}$ handfuls in this case.

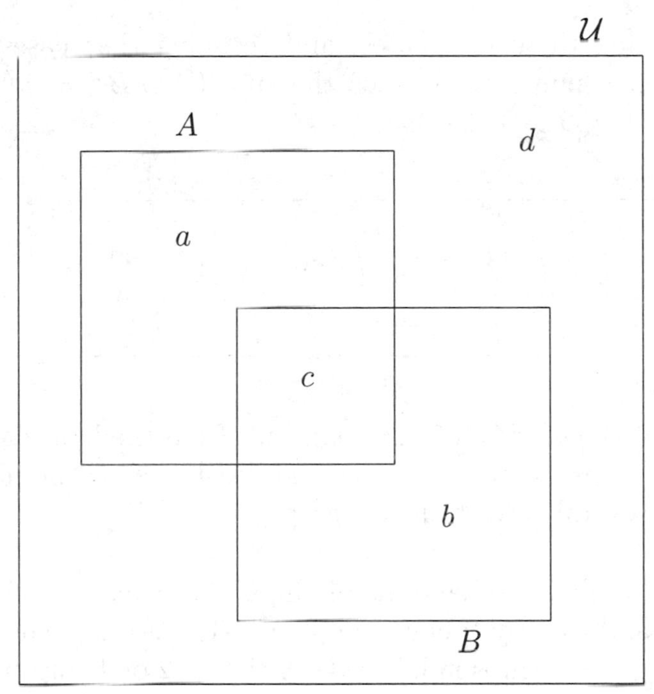

2. A 5-chip handful with exactly 1 chip numbered 12 and no chips numbered 13 is formed as follows. Choose 1 of 4 chips numbered 12 in 4 ways. Then choose 4 chips from a set of 44 chips, (ignoring those chips numbered with 12 and 13), in $\binom{44}{4}$ ways. Thus, there are $4 \cdot \binom{44}{4}$ handfuls in this case.

3. A 5-chip handful with no chips numbered 12 and exactly 1 chip numbered 13 is the same problem we solved in part 2. Thus, there are $4 \cdot \binom{44}{4}$ such handfuls.

4. A 5-chip handful with exactly 1 chip numbered 12 and exactly 1 chip numbered 13 is formed as follows. Choose 1 of 4 chips

numbered 12 in 4 ways, choose 1 of 4 chips numbered 13 in 4 ways, and then choose 3 chips from the set of 44 chips that do not include 12 or 13 in $\binom{44}{3}$. Thus, there are $4^2 \cdot \binom{44}{3}$ handfuls in this case.

Combining these four cases, and recalling that cases 2 and 3 give the same number, gives us the sum $(A \cup B)' = n(\text{case } 1) + 2n(\text{case } 2) + n(\text{case } 4)$ so that

$$d = \binom{44}{5} + 8 \cdot \binom{44}{4} + 4^2 \cdot \binom{44}{3}.$$

$A \cap B$ is the set of all 5-chip handfuls from the bag that contain at least 2 chips numbered 12 and at least 2 chips numbered 13. There are several cases to investigate.

1. In this case the handfuls of chips have exactly 2 chips numbered 12, and exactly 2 chips numbered 13. Form such a handful by choosing 2 of 4 chips numbered 12, choose 2 of 4 chips numbered 13, and then choose 1 of the remaining 44 chips. Thus, there are $44 \cdot \binom{4}{2}^2$ such handfuls in this case.

2. These handfuls consist of exactly 3 chips numbered 12 and exactly 2 chips numbered 13. Choose 3 of 4 chips numbered 12 in $\binom{4}{3} = 4$ ways, and choose 2 of 4 chips numbered 13 in $\binom{4}{2}$ ways. These choices fill out the handful. There are $4 \cdot \binom{4}{2}$ such handfuls in this case.

3. Similarly, there are $4 \cdot \binom{4}{2}$ handfuls of exactly 2 chips numbered 12 and exactly 3 chips numbered 13.

Then $n(A \cap B) = n(\text{case } 1) + 2n(\text{case } 2)$ so that

$$c = 44 \cdot \binom{4}{2}^2 + 8 \cdot \binom{4}{2}$$

As usual, $n(A \cap B') = n(A' \cap B)$, so by looking at the Venn diagram we see that

$$\begin{aligned} n(\mathcal{U}) &= n(A \cup B)' + n(A \cap B) + 2n(A \cap B') \\ n(A \cap B') &= \frac{1}{2}\left(n(\mathcal{U}) - n(A \cup B)' - n(A \cap B)\right) \end{aligned}$$

Substituting our above work into the equation yields

$$\begin{aligned} a = n(A \cap B') = \frac{1}{2}\Bigg(\binom{52}{5} \\ - \binom{44}{5} - 8 \cdot \binom{44}{4} - 4^2 \cdot \binom{44}{3} \\ -44 \cdot \binom{4}{2}^2 - 8 \cdot \binom{4}{2}\Bigg) \end{aligned}$$

This completes the Venn diagram.

The last region $A \cap B'$ of the Venn diagram in the above example can also be filled in using a case-by-case analysis as follows.

The set $A \cap B'$ is the set of 5-chip handfuls that contain at least 2 chips numbered 12 and that do not contain at least 2 chips numbered 13. That is, $(A \cap B')$ is the set of 5-chip handfuls that contain at least 2 chips numbered with 12, and that contain at most 1 chip numbered with 13. The cases are

exactly 2 chips numbered 12 and no chips numbered 13
exactly 3 chips numbered 12 and no chips numbered 13
exactly 4 chips numbered 12 and no chips numbered 13
exactly 2 chips numbered 12 and exactly 1 chip numbered 13
exactly 3 chips numbered 12 and exactly 1 chip numbered 13
exactly 4 chips numbered 12 and exactly 1 chip numbered 13

Since there are only 4 chips numbered 12 in the bag and since the handfuls in $A \cap B'$ contain at most 1 chips numbered 13, these are all of the possible cases. If you choose to follow this line of thought, then remember that in addition to 12's and 13's you must fill a handful of 5-chips. I prefer the method used to find $n(A \cap B')$ in the example.

12.2 Exercises

The 5-chip handfuls are from our bag given in Notation 12.1.1.

1. Count the number of 5-chip handfuls consisting of at least 3 red round chips.

Answer: $\binom{52}{5} - \binom{13}{2}\binom{39}{3} - 13 \cdot \binom{39}{4} - \binom{39}{5}$

If you got the **wrong** answer: $\binom{13}{3}\binom{52}{2}$ then your construction of the handful is incorrect.

2. Count the number of 5-chip handfuls consisting of at most 1 red round chips and at most 1 black round chips.

Answer: $\binom{26}{5} + 2 \cdot 13 \cdot \binom{26}{4} + 13^2 \binom{26}{3}$

In exercises 3 through 5, Choose a 5-chip handful from our bag.

3. Say the handful consists of at least 1 red round chip or at least 1 black round chip. Label the associated Venn diagram.

4. Say the handful consists of exactly 3 red round chips or exactly 2 black round chips. Label the associated Venn diagram.

Answer: $c = 1$, $a = \binom{13}{3}\left(\binom{26}{2} + 13 \cdot 26 + \binom{13}{2}\right)$,

$b = \binom{13}{2}\left(\binom{26}{3} + 13 \cdot \binom{26}{2} + 26 \cdot \binom{13}{2} + \binom{13}{3}\right)$,

$d = 26^3 - a - b - c$.

5. Say the handful consists of at least 2 red round chips or at least 2 black round chips. Label the associated Venn diagram.

Chapter 13

Poker and Counting

Let us use the counting methods from previous chapters to solve some problems about poker. Note: We are not counting cards in the manner that that was made popular in a movie filmed in the late twentieth century. We are counting the number of poker hands of a specific type. Nothing that we are doing here will aid anyone in playing the game of five card poker. But I find the counting to be quite some fun.

A *standard deck* consists of 52 cards labeled with the 13 values $2, 3, 4, 5, 6, 7, 8, 9, 10, J, Q, K, A$. These labels are called *kinds of cards*. Each kind of card comes in 4 *suits*. These suits are *hearts*, *diamonds*, *clubs*, and *spades*. Let $\mathcal{U}$ be the set of 5 card hands (poker hands) from the standard deck of 52 cards. Then

$$n(\mathcal{U}) = \binom{52}{5}.$$

The kinds of cards have *value*. For instance, 3 has more value than 2, 4 has more value than 3, and so on. The *Ace* has more value than the other cards. A *poker hand* is a hand of 5 cards from a standard deck.

13.1 Warm-Up Problems

Let us count a few important poker hands that do not directly influence our counting of all poker hands.

EXAMPLE 13.1.1 Count the number of poker hands that contain at most 1 Q or at most 1 K.

Solution: The set A contains poker hands that contain at most 1 Q, and the set B contains poker hands that contain at most 1 K. We are asked to count $n(A \cup B)$. To do so we will use the complement law.

Note that

$$n(A \cup B) = n(\mathcal{U}) - n(A \cup B)' = n(\mathcal{U}) - n(A' \cap B')$$

and that $A' \cap B'$ is the set of poker hands from the 44 cards other than Q and K. There are $\binom{44}{5}$ such hands, so that $n(A \cup B) =$

> The number of poker hands of at most 1 Q or at most 1 K
>
> $$= \binom{52}{5} - \binom{44}{5}$$

is our count.

EXAMPLE 13.1.2 Count the number of poker hands that contain at least 2 Q's and at least 2 K's.

Solution: In order to avoid double counting, we will take cases for that fifth card. Then a poker hand that contains at least 2 Q's and at least 2 K's is formed as in the following two cases.

CASE 1: Choose 2 of the 4 Q's and 2 of the 4 K's in $\cdot \binom{4}{2}^2$ ways. Then choose 1 of the 44 cards that is not a Q or a K in 44 ways.

CASE 2: Choose 3 of the 4 Q's, and then 2 of the 4 K's in $4 \cdot \binom{4}{2}$ ways. The same number occurs if we reverse the roles of Q and K in Case 2.

Then

> The number of hands of at least 2 Q's and at least 2 K's
> $$= 44 \cdot \binom{4}{2}^2 + 2 \cdot 4 \cdot \binom{4}{2}$$

EXAMPLE 13.1.3 Count the number of poker hands containing at least 2 Q's or at least 2 K's.

Solution: Let A be the set of poker hands that contain at least 2 Q's, and let B be the set of poker hands that contain at least 2 K's. By the inclusion/exclusion principle we have

$$n(A \cup B) = n(A) + n(B) - n(A \cap B).$$

The number $n(A)$ is calculated using the complement law,

$$n(A) = n(\mathcal{U}) - n(A')$$

where A' contains at most 1 Q. A hand in A' has exactly 1 or no Q's.

Construct a hand in A' as follows. Choose 1 Q in exactly 4 ways, choose 4 cards (respectively, 5 cards) that are not a Q in 48 ways. Then

$$n(B') = n(A') = \binom{48}{5} + 4 \cdot \binom{48}{4}$$

so that

$$n(B) = n(A) = \binom{52}{5} - \binom{48}{5} - 4 \cdot \binom{48}{4}.$$

We calculated $n(A \cap B)$ in Example 13.1.2 as $44 \cdot \binom{4}{2}^2 + 2 \cdot 4 \cdot \binom{4}{2}$, so that

The number of poker hands of at least 2 Q's or at least 2 K's

$$= 2 \cdot \left(\binom{52}{5} - \binom{48}{5} - 4 \cdot \binom{48}{4} \right) - \left(44 \cdot \binom{4}{2}^2 + 8 \cdot \binom{4}{2} \right)$$

This completes our count.

13.2 Poker Hands

A *flush* is a poker hand consisting of cards of one suit. There are flushes of hearts, flushes of diamonds, flushes of clubs, and flushes of spades. And that is all of them.

EXAMPLE 13.2.1 Count the number of flushes.

Solution: To draw a flush, you first choose the 1 of 4 suits to be used in the flush. Then you choose the 5 of 13 cards of that suit. Thus, there are

$$4 \cdot \binom{13}{5}$$

flushes in the game of poker.

A *straight* is a poker hand that contains 5 cards in sequence. For instance, a hand that contains five cards of the kinds $2, 3, 4, 5, 6$ is a straight as is $10, J, Q, K, A$. We will not include $A, 2, 3, 4, 5$ as a straight.

EXAMPLE 13.2.2 Count the number of straights.

Solution: The lowest kind of card to begin a straight is a 2, and the highest kind of card to begin a straight is a 10.

We can construct a straight by choosing one of the 9 kinds of cards 2, 3, 4, 5, 6, 7, 8, 9, 10 that will begin the sequence of five cards. The other 4 kinds of cards are automatically chosen to fill out the 5 card straight. Then for each of these 5 kinds of cards we pick 1 of the 4 cards of that kind in exactly 4^5 ways. Then the number of straights in poker is

$$9 \cdot 4^5.$$

EXAMPLE 13.2.3 Count the number of poker hands that are either straights or flushes.

Solution: Let F be the set of flushes and let S be the set of straights. By the inclusion/exclusion principal the number of poker hands that are flushes or straights is

$$n(F \cup S) = n(F) + n(S) - n(F \cap S).$$

From Examples 13.2.1 and 13.2.2 we count

$$n(F) = 4 \cdot \binom{13}{5} \text{ and } n(S) = 9 \cdot 4^5.$$

The intersection $F \cap S$ is the set of *straight flushes*, (straights of the same suit). To count straight flushes, begin by choosing the suit the flush is to have. There are 4 choices. Then choose the beginning of your straight in 9 ways, as we did in Example 13.2.2. Within the chosen suit there is only one choice for each of the 5 cards in the straight flush. Thus, there are $n(F \cap S) = 4 \cdot 9$ straight flushes. Then

The number of poker hands that are straights or flushes

$$= 4 \cdot \binom{13}{5} + 9 \cdot 4^5 - 4 \cdot 9.$$

To this point we have counted the number of poker hands based on straights and flushes. We will use what we have learned from the first section of this chapter and start to count poker hands that contain at least two cards of a kind. There are several cases to consider. In our counting of these hands notice that we will not recognize any value in these hands, just arrangements of five cards.

EXAMPLE 13.2.4 Count the poker hands that are exactly 1 pair.

Solution: We choose the kinds of cards X, U, V, W so that X will be the kind of the pair, U, V, W will be cards different from X and from each other. Count the pairs of kind X by choosing 1 of 13 kinds and then choosing 2 of 4 X's from the deck. Do this in exactly $13 \cdot \binom{4}{2}$ ways. Next, from the remaining 12 kinds, choose the three kinds U, V, W in $\binom{12}{3}$, and then for each of these kinds choose 1 of the 4 of that kind. This is done in 4^3 ways. Hence

The number of poker hands of exactly 1 pair

$$= 4^3 \cdot 13 \cdot \binom{4}{2} \cdot \binom{12}{3}$$

In the next couple of examples we count the number of hands that are either two pair or a full house. Let U, V, W, X be unequal kinds of cards. If we ignore the suits being used, then a hand of exactly a pair of X's can be written as X, X, U, V, W and a hand of two pair can be written as X, X, U, U, V.

EXAMPLE 13.2.5 Let X, Y, Z be unequal kinds of cards. Count the number of poker hands of the form

1. X, X, Y, Y, Z, or
2. X, X, X, Y, Y.

Solution: 1. Count the number of poker hands X, X, Y, Y, Z. Choose unequal kinds X and Y from the set of 13 kinds of cards in $\binom{13}{2}$ ways, and then choose 2 of the 4 cards of the kind X and then choose 2 of the 4 cards of the kind Y in a total of $\binom{4}{2}^2$ ways. Choose Z from the 44 cards that are not X or Y in 44 ways. Thus, the number of hands X, X, Y, Y, Z is

$$n(X, X, Y, Y, Z) = 44 \binom{13}{2} \binom{4}{2}^2.$$

2. Suppose that the hand is X, X, Y, Y, Y. Choose a pair of X's in $13 \cdot \binom{4}{2}$ ways, and choose three Y's in $12 \cdot \binom{4}{3} = 12 \cdot 4$ ways. Thus, the number of hands in this case is

$$n(X, X, X, Y, Y) = 4 \cdot 13 \cdot 12 \cdot \binom{4}{2}.$$

This completes the proof.

EXAMPLE 13.2.6 1. By Example 13.2.5(1),

$$n(\text{exactly 2 pair}) = 44 \binom{13}{2} \binom{4}{2}^2.$$

2. By Example 13.2.5(2),

$$n(\text{a full house}) = 13 \cdot 12 \cdot 4 \cdot \binom{4}{2}.$$

3. Hence

The number of poker hands that are two pair or a full house

$$= 44 \cdot \binom{13}{2} \binom{4}{2}^2 + 13 \cdot 12 \cdot 4 \cdot \binom{4}{2}$$

EXAMPLE 13.2.7 1. Count the number of poker hands that are exactly three of a kind.

Solution: Let us again choose three unequal kinds X, Y, Z of cards. If we ignore the suits used, then a hand of exactly three of a kind can be written as X, X, X, U, V. Such a hand is constructed by first choosing the kind X and the 3 cards from the 4 of kind X in $13 \cdot \binom{4}{3}$ ways. Choose 2 kinds from the remaining 12 kinds in $\binom{12}{2}$ ways, and then choose the cards U and V in 4^2 ways. Hence

The number of poker hands of exactly 3 of a kind

$$= 4^2 \cdot 13 \cdot \binom{4}{3}\binom{12}{2}.$$

2. Count the number of poker hands that are exactly four of a kind.

Solution: The required hand X, X, X, X, U is constructed by first choosing the kind X in 13 ways. Choose the cards of kind X in 1 way. Choose one of the remaining 12 kinds in 12 ways, and then choose the card U in 4 ways. Then

The number of poker hands of exactly 4 of a kind

$$= 4 \cdot 12 \cdot 13.$$

It remains to count the hands that are less than a pair of deuces. These are the hands that have *no real value* in poker. They will lose to anything except a possible high card. These are the hands that are neither flushes nor straights.

EXAMPLE 13.2.8 Count the number of poker hands with value less than a pair.

Solution: We cannot allow pairs, so we choose 5 different kinds of cards U, V, W, X, Y for our hand in $13 \cdot 12 \cdot 11 \cdot 10 \cdot 9$ ways. Now choose 1 of the 4 cards for each of kinds U, V, W, X in 4^4 ways. The reason we choose in this way is that we wish to avoid a flush.

CASE 1: If the four cards chosen are of 1 suit s, then fill in the hand by choosing the fifth card from 1 of the 3 cards of kind Y that are not of the suit s.

CASE 2: Otherwise, a flush is not possible, and so we choose 1 of the 4 of kind Y in 4 ways.

Thus, $n(U, V, W, X, Y) =$

$$3 \cdot 4^4 \cdot 13 \cdot 12 \cdot 11 \cdot 10 \cdot 9 + 4 \cdot 4^4 \cdot 13 \cdot 12 \cdot 11 \cdot 10 \cdot 9 - n(\text{straight not flush}),$$

and in the manner that we counted in Examples 13.2.3 we see that $n(\text{straight not flush}) = 4^5 \cdot 9 - 4 \cdot 9$. Hence

The number of poker hands of value less than a pair

$$= 7 \cdot 4^4 \cdot 13 \cdot 12 \cdot 11 \cdot 10 \cdot 9 - 4^5 \cdot 9 + 4 \cdot 9$$

At last we are ready to count the number of different poker hands from a standard deck of 52.

EXAMPLE 13.2.9 Count the number of 5 card poker hands from a standard deck of 52.

Solution: A hand in poker is has exactly 1 pair, or exactly 2 pair, or exactly 3 of a kind, or exactly a full house, or exactly 4 of a kind, or a straight, or a flush, or a straight flush, or it has no value. Hands containing a multiple number of cards of 1 kind are counted in Examples 13.2.3, 13.2.4, 13.2.6, 13.2.7, and 13.2.8. So as to organize the information, we list the solutions given in these examples in a chart.

Straights or flushes	$4 \cdot \binom{13}{5} + 9 \cdot 4^5 - 4 \cdot 9$
1 Pair	$4^3 \cdot 13 \cdot \binom{4}{2} \cdot \binom{12}{3}$
2 Pair	$44\binom{13}{2}\binom{4}{2}^2$
Full House	$13 \cdot 12 \cdot 4 \cdot \binom{4}{2}$
3 of a Kind	$4^2 \cdot 13 \cdot \binom{4}{3}\binom{12}{2}$
4 of a Kind	$4 \cdot 12 \cdot 13$
No value	$7 \cdot 4^4 \cdot 13 \cdot 12 \cdot 11 \cdot 10 \cdot 9$ $-4^5 \cdot 9 + 4 \cdot 9$

The sets counted by the rows in the above chart are disjoint. No two of them contain a common hand. For instance, we have counted the hands in such a way that the hand X, X, U, V, W and the hand X, X, Y, Y, Z are counted by different lines in the chart. In particular, n(a pair and two pair) $= 0$. So the inclusion/exclusion principle implies that

$$\begin{aligned} &n(\text{a pair or two pair}) \\ &\qquad = n(\text{a pair}) + n(\text{two pair}) \text{ - } n(\text{a pair and two pair}) \\ &\qquad = n(\text{a pair}) + n(\text{two pair}) \end{aligned}$$

there is no need to compensate for double counting when counting full houses and three of a kind.

Therefore,

The number of different 5 card poker hands

$$= 4 \cdot \binom{13}{5} + 4^3 \cdot 13 \cdot \binom{4}{2} \cdot \binom{12}{3} + 44 \binom{13}{2} \binom{4}{2}^2 + 13 \cdot 12 \cdot 4 \cdot \binom{4}{2} + 4^2 \cdot 13 \cdot \binom{4}{3} \binom{12}{2} + 4 \cdot 12 \cdot 13 + 7 \cdot 4^4 \cdot 13 \cdot 12 \cdot 11 \cdot 10 \cdot 9.$$

This completes our count of the distinct poker hands.

13.3 Jacks or Better

To make future counting work, identify J with 11, Q with 12, K with 13, and A with 14. Identify the numbered cards 2-10 with their numbers. Thus, a 10 is identified with 10 and A is identified with 14. Then given a card U there are $15 - U$ cards on the list $U, \ldots, A$. For example, if $U = J$ then identify U with 11, and there are $15 - 11 = 4$ cards on the list J, Q, K, A.

Fix a kind of card X. *A pair of X's* is a poker hand containing exactly 2 X's and 3 other unequal kinds of cards U, V, W that are different from X. A hand is said to be *at least a pair of X's* if its value as a poker hand is a pair of X's or better. For instance, a poker hand $2, 3, 3, 4, 5$ would be *at least a pair of 3's*, as would the straight $2, 3, 4, 5, 6$. Our goal is to count the number of hands that are J's or better.

EXAMPLE 13.3.1 Let U be a number associated with the kind U. Count the number of poker hands that are no more than a pair of U's.

Solution: A hand that is no more than a pair of U's is at most a pair of the $U - 2$ kinds whose number value X satisfies $X < U$, or it has no value. Then, as in Example 13.2.4,

$$n(\text{a pair of } X\text{'s such that } X < U) = 4^3 \cdot (U-2) \cdot \binom{4}{2} \cdot \binom{U-3}{3},$$

and by Example 13.2.8,

$$n(\text{hands with no value}) = 7 \cdot 4^4 \cdot 13 \cdot 12 \cdot 11 \cdot 10 \cdot 9 - 4^5 \cdot 9 + 4 \cdot 9.$$

Hence

> The number of poker hands that are no more than a pair of U's
>
> $$= 4^3 \cdot (U-2) \cdot \binom{4}{2} \cdot \binom{U-3}{3} + 7 \cdot 4^4 \cdot 13 \cdot 12 \cdot 11 \cdot 10 \cdot 9 - 4^5 \cdot 9 + 4 \cdot 9.$$

This completes the count.

The number of poker hands counted in Example 13.3.1 that are no more than a pair of U's is denoted by the parameterized symbol

$$n(\text{PH}(U))$$

and we let

$$n(\text{PH})$$

denote the number of different kinds of poker hands that can be dealt from the standard 5-card deck found on page 159.

EXAMPLE 13.3.2 Count the number of poker hands that are U's or better.

Solution: A hand of U's or better is complemented by a hand that is no more than U, so that the number of poker hands that are U's or better $= n(\text{PH}) - n(\text{PH}(U))$. The number $n(\text{PH})$ is found in Example 13.2.9, and the number $n(\text{PH}(U))$ is found in Example

13.3.1. Hence

The number of poker hands that are U's or better

$$= 4 \cdot \binom{13}{5} + 4^3 \cdot 13 \cdot \binom{4}{2} \cdot \binom{12}{3} + 44 \binom{13}{2} \binom{4}{2}^2$$
$$+ 13 \cdot 12 \cdot 4 \cdot \binom{4}{2} + 4^2 \cdot 13 \cdot \binom{4}{3} \binom{12}{2} + 4 \cdot 12 \cdot 13.$$
$$- 4^3 \cdot (U-2) \cdot \binom{4}{2} \cdot \binom{U-3}{3} + 4^5 \cdot 9 - 4 \cdot 9.$$

This completes our count.

EXAMPLE 13.3.3 Count the number of poker hands that are J's or better.

Solution: The number equivalent of J is $U = 11$, so by Example 13.3.2,

The number of poker hands that are J's or better

$$= 4 \cdot \binom{13}{5} + 4^3 \cdot 13 \cdot \binom{4}{2} \cdot \binom{12}{3} + 44 \binom{13}{2} \binom{4}{2}^2$$
$$+ 13 \cdot 12 \cdot 4 \cdot \binom{4}{2} + 4^2 \cdot 13 \cdot \binom{4}{3} \binom{12}{2} + 4 \cdot 12 \cdot 13.$$
$$- 4^3 \cdot 9 \cdot \binom{4}{2} \cdot \binom{8}{3} + 4^5 \cdot 9 - 4 \cdot 9.$$

This completes our count.

13.4 Exercises

A poker hand has 5 cards from the standard deck.

1. How many poker hands have exactly 2 pair?

Answer: $\binom{13}{2}\binom{4}{2}^2 \cdot 44$.

Wrong answer: $13 \cdot 12 \cdot \binom{4}{2}^2 \cdot 44$. When you count poker hands you must assume that the 5 cards came to you at once. Any other method of counting must agree with this condition. Order in the hand does not matter so you choose the 2 kinds in $\binom{13}{2}$ ways.

2. How many poker hands have exactly 3 of 1 kind and 2 of another kind? Answer: $13 \cdot 12 \binom{4}{3}\binom{4}{2}$ Choose a kind for the 3 of a kind and then choose a kind for the 2 of a kind in $13 \cdot 12$ ways.

This applies to exercises 3 and 4. Suppose you are playing poker and you hold a hand $2, 3, 6, 9, 10$. You throw away the 2 and the 3 and you pray for a 7 and an 8 to fill in a straight. You get 2 new cards. Let A be the set of hands in which you have a 7, and let B be the hand in which you have an 8.

3. Describe $A \cap B$? Answer: You hold $6, 7, 8, 9, 10$. The players tell me this is called *filling an inside straight.*

4. Fill in the associated Venn diagram.

Suppose you are playing poker and you hold a hand $2, 3, 7, 8, 10$. You throw away the 2 and the 3 and you pray for 2 cards to fill in a straight. This is done with a 6 and a 9, or a 9 and a J. You get 2 new cards. Let A be the set of hands in which you have a 6, let B be the hand in which you have an 9, let C be the hand in which you have a J.

5. Describe $A \cap B$? Answer: You hold $6, 7, 8, 9, 10$.
6. Describe $A \cap C$? Answer: You hold $6, 7, 8, 10, J$.
7. Describe $B \cap C$? Answer: You hold $7, 8, 9, 10, J$.
8. Fill in the associated Venn diagram.

Chapter 14

Advanced Counting

The counting problems involving permutations and combinations are predicated on the fact that the things being counted are different. If the letters in the word being counted are repeated, the places they occupy are different. When choosing subsets of a set, the elements of the set are different. When choosing chips from a bag or cards from a deck, the chips and cards are different. We will try to count in some problems where the objects are indistinguishable.

14.1 Indistinguishable Objects

In the first example we will discuss the problem of choosing *in distinguishable* objects from a page. These objects exist on your typewriter, in a mathematical universe called a Platonic Universe of shapes and ideas. They cannot be touched, and once chosen you cannot distinguish one from the other. They exist in a container of some kind, called a *page*. That page could be a sheet of paper, or it could be a container of some kind. Our indistinguishable objects will be the letter A typed over many times on a sheet of paper. Of course we will also use B and C in the same way. For example, the page of indistinguishable A's could be $AAAAAAAAAA$. This might do for one problem, or you might fill a larger (abstract mathematical) typed page with A's. This page extends for as far as you need it. Such a page exists in our Platonic mathematical universe.

In our mathematical setting, *indistinguishable* means the following. Observe one A on the page and then observe another A. You cannot tell the difference between these A's. Physical location is not noted when you choose your A. You choose a collection of A's by randomly pointing to some set of them on the page. Notice that you do not remove anything from the page. You see your choices, but you do not handle them.

This is very different from the chips that we chose in previous chapters, as those chips were different in shape, color, and numbered label. You chose them from a bag. The cards we worked with were also different elements, and you chose them from the deck. Our chosen A's are just what we see on the page.

Some of the arguments used in this chapter are *indirect arguments.* That is, we begin with a premise. We begin by *assuming that we cannot do something*, and then proceed to argue. The argument will lead us to a mathematical mistake called a *contradiction.* This contradiction is called a *Falsehood.* It is untrue. We began with a premise that we could not do something, we argued, and then we deduced a falsehood. The only way that this can happen is when the premise is incorrect. We conclude, then, that *we can do that something.*

EXAMPLE 14.1.1 Suppose that we have a page containing two identical A's. In how many different ways can we choose 1 A from the page?

Solution: The answer is just 1. There is just 1 way, with replacement, to choose 1 A from the page of 2 identical A's. Here is why.

The argument is indirect. Suppose, to the contrary, that we can choose 1 A in 2 different ways. We search for a contradiction.

These 2 choices are different. That is, our choices of A's are different, which does not agree with our assumption that the A's are indistinguishable. And this is the contradiction that we were looking for. Thus, we began with the false premise that there are 2 different ways to choose A's. We conclude that there is only 1 way to choose an A from the page.

This might come as an academic shock to you. There should be more than 1 way to choose 2 A's, you might say. You forget that these are indistinguishable A's and that they are left on a page after you choose them.

In other words, when you make that choice of an A and then another A, making that choice as I describe above, you have no way to tell if you chose the first A twice or if you chose 2 A's in different locations on the page. Remember, location is not observed when you choose A's from the page. Since you cannot make a distinction between chosen A's you can only make one choice of A from the page.

Let us investigate the consequences of this type of choosing.

EXAMPLE 14.1.2 Let $n \geq k > 0$ be whole numbers. Let us take a page of exactly n identical A's. In how many ways can you choose k A's from a page of n identical A's.

Solution: We will show that there is exactly 1 way to make this choice. Let us argue indirectly. Suppose, to the contrary, that you can make 2 choices C and D of exactly k A's from the page. (That is, just look at the page and see k A's.) Since C and D are different, the A's in C are different from the A's in D. This contradicts our hypothesis that the A's on the page are indistinguishable. Hence, we began with a false premise. Our assumption that there are at least 2 different ways to choose k A's is false. We conclude that there is only 1 way to choose k A's from n identical A's.

Thus, there is exactly 1 way to choose 250 A's from a page of 525 identical A's.

EXAMPLE 14.1.3 If there are $k = 0$ identical A's to be chosen and if we find two choices C and D containing 0 A's then C and D have 0 A's in them, so $C = D = \emptyset$. Thus, there is exactly 1 way to choose $k = 0$ identical A's from the page. That is, C and D are the same choice of $k = 0$ A's from the page. Thus, there is exactly 1 way to choose no A's from a page of identical A's.

EXAMPLE 14.1.4 Suppose that there are exactly n indistinguishable A's on a page. Show that there are $n + 1$ ways you can choose a collection of A's from this page.

Solution: Make a choice of A's from the page. Your choice has exactly k A's for some whole number $0 \leq k \leq n$, and by Examples 14.1.2 and 14.1.3, there is only 1 way to choose k A's from the page. That is, the number of A's in your choice completely determines your choice. Thus, there is exactly 1 choice of 1 A's, there is exactly 1 choice of 2 A's, $\cdots$, there is exactly 1 choice of n A's, and since we allow the most trivial of choices, there is exactly 1 choice of no A's. Adding up these numbers we see that there are $n + 1$ choices of A's from a page of n indistinguishable A's.

For example, there are 3 ways to choose A's from the page AA of 2 indistinguishable A's. Check this for yourself.

Let us make the problem a little more interesting by introducing the multiplication principal.

EXAMPLE 14.1.5 There are n indistinguishable A's and m indistinguishable B's on a page. In how many ways can you choose a collection of A's and B's?

Solution: There are two tasks. Choose a collection of identical A's, which by Example 14.1.4 can be done in $n+1$ ways, and choose a collection of identical B's from the page, which can be done in $m+1$ ways. Then by the multiplication principal there are $(n+1)(m+1)$ total choices of A's and B's from a the page of identical A's and identical B's.

For instance, if we are to count the number of ways we can choose a collection of A's and B's from a page of 10 indistinguishable A's and 11 indistinguishable B's then the number of choices is $10 \cdot 11 = 110$ collection. Compare this example to the next example.

EXAMPLE 14.1.6 There are 4 indistinguishable A's and 5 indistinguishable B's on a page. In how many ways can you choose 3 letters from the page?

Solution: There are four cases to consider.

1. No B's are chosen. Then we are counting the number of choices of 3 A's from the page. By Example 14.1.2, there is 1 way to make this choice.

2. Exactly 1 B is chosen. Then we choose 2 A's from the page. By Example 14.1.2, there is 1 way to make this choice.

3. Exactly 2 B's are chosen. The we choose 1 A from the page. By Example 14.1.2, there is 1 way to make this choice.

4. Exactly 3 B's are chosen. Then we choose no A's. By Example 14.1.2, there is 1 way to make this choice.

The total number of choices of 3 letters from the page of indistinguishable A's and indistinguishable B's from this page is then

$$1 + 1 + 1 + 1 = 4.$$

14.2 Circular Permutations

Assume there is a rather large circular table, complete with chairs. You have a number of people you wish to seat at the table. In how many different ways can you seat these people at the table? To answer this we must know the following. *What does it mean for 2 seatings at this table to be the same?* We abstract this problem without losing the spirit of the problem. A *label* of the unit circle is an n tuple $(x_1, \ldots, x_n)$ of the different numbers from $\{1, \ldots, n\}$. Then $(2, 1)$ and $(3, 2, 1)$ are labels of the unit circle. You can physically envision this labeling by equally spacing n points on the unit circle and then assigning clockwise the whole number x_i to point i. This describes the label $(x_1, \ldots, x_n)$. Thus, in the labeling $(4, 3, 2, 1)$ of the unit circle, 4 labels 1, 3 labels 2, 2 labels 3, and 1 labels 4.

We ask the simple question *What does it mean for 2 such labels of the unit circle to be the same?*

First we need to know what *same* means in this context. We will say that 2 labels $(x_1, \ldots, x_n)$ and $(y_1, \ldots, y_m)$ of the unit circle are *the same* if there is a whole number $1 \leq k \leq n$ such that

$$(x_k, \ldots, x_n, x_1, \ldots, x_{k-1}) = (y_1, \ldots, y_m).$$

From this definition, it is clear that if 2 labels $(x_1, \ldots, x_n)$ and $(y_1, \ldots, y_m)$ are the same then they have the same number of terms. That is, $n = m$. Two labels are *different* if they are not the same.

Visually, and imprescisely, we say that one label X of the unit circle is the same as another label Y of the unit circle if we can rotate the circle and labels so that the numbers in X match up with the same numbers in Y.

Paraphrasing, 2 labels $(x_1, \ldots, x_n)$ and $(y_1, \ldots, y_m)$ are the *same* if you can cut off some initial piece of $(x_1, \ldots, x_n)$, place it at the end of $(x_1, \ldots, x_n)$, and arrive at $(y_1, \ldots, y_m)$. Hence, the 2 labels of the unit circle differ only in where they start. The order of the labeled points is the same in both labels.

For instance, the labels $(1, 2, 3)$ and $(2, 3, 1)$ are the same, because we can cut off the 1, making $(2, 3)$, and then placing 1 the end of it this way $((2, 3), 1) = (2, 3, 1)$ which is the $(2, 3, 1)$ we wanted.

Let us write down the labels with 4 numbers that are the same as $(4, 2, 3, 1)$. These are formed by cutting initial piece of $(4, 3, 2, 1)$ and placing it at the end of $(4, 3, 2, 1)$. Observe. Given $(4, 3, 2, 1)$, cut off the first number 4 and place it at the end as in $(3, 2, 1, 4)$. So $(4, 3, 2, 1)$ is the same as $(3, 2, 1, 4)$. Cut off $(4, 3)$ from $(4, 3, 2, 1)$ and place it on the end to form $(3, 2, 4, 1)$. Thus, $(4, 3, 2, 1)$ is the same as $(3, 2, 4, 1)$. Similarly, cut $(4, 3, 2)$ from $(4, 3, 2, 1)$ and place it at then end to show that $(4, 3, 2, 1)$ is the same as $(1, 4, 3, 2)$.

Observe that $(1, 2, 3)$ and $(1, 3, 2)$ are not the same. Work with this example for a while before continuing.

EXAMPLE 14.2.1 Let $\mathbf{x} = (x_1, \ldots, x_n)$ be a label of the unit circle. Then $\mathbf{x}$ is the same as n labels.

Solution: We write down the labels that are the same as $\mathbf{x}$ by systematically placing initial parts of $\mathbf{x}$ at the end of $\mathbf{x}$ as follows.

$$\begin{gathered}
(x_1, \ldots, x_n) \text{ is the same as } \mathbf{x}, \\
(x_2, \ldots, x_n, x_1) \text{ is the same as } \mathbf{x}, \\
(x_3, \ldots, x_n, x_1, x_2) \text{ is the same as } \mathbf{x}, \\
\vdots \\
(x_n, x_1, \ldots, x_{n-1}) \text{ is the same as } \mathbf{x}.
\end{gathered}$$

By the definition of the word *same*, this is a list of all of the labels that are the same as $\mathbf{x}$. Thus, there are n labels that are the same as $\mathbf{x}$.

Now that we know how many labels are the same, we ask for the number of different labels of the unit circle with n numbers.

EXAMPLE 14.2.2 In how many ways can you label the unit circle with 3 numbers?

Solution: The problem is to count the number of different labels (x_1, x_2, x_3).

A label is a permutation of the numbers $\{1, 2, 3\}$. The number of such permutations is 3!.

Let L be the number of different labels $\mathbf{x} = (x_1, x_2, x_3)$. (Recall our definition of *different*.) Write down the L different labels as

$$\mathbf{x}_1, \ldots, \mathbf{x}_L.$$

By Example 14.2.1, each $\mathbf{x}_i$ is the same as exactly 3 labels $\mathbf{x}_i, \mathbf{y}_i, \mathbf{z}_i$. We then have an array

$$\begin{array}{ccc} \mathbf{x}_1 & \cdots & \mathbf{x}_L \\ \mathbf{y}_1 & \cdots & \mathbf{y}_L \\ \mathbf{z}_1 & \cdots & \mathbf{z}_L \end{array}$$

consisting of 3 rows and L columns. Moreover, the labels in this array account for all of the 3! labels (permutations) of the numbers $\{1, 2, 3\}$. Hence

$$3! = 3L$$

so that the number of different labels is $L = 3!/3 = 2$.

The general problem is solved in the same way.

EXAMPLE 14.2.3 There are $(n-1)!$ labels of the unit circle with n numbers.

Solution: A label is a permutation $(x_1, \ldots, x_n)$ of $\{1, \ldots, n\}$. There are $n!$ such permutations.

Let L be the different labels of the unit circle using n numbers. List these different labels.

$$\mathbf{x}_1, \mathbf{x}_2, \ldots, \mathbf{x}_L$$

Pick a label $\mathbf{x}_i$. By Example 14.2.1, $\mathbf{x}_i$ is the same as exactly n labels. Write out the labels that are the same as $\mathbf{x}_i$.

$$\begin{array}{c} \mathbf{x}_{i1} \\ \vdots \\ \mathbf{x}_{in} \end{array}$$

Put these columns together to form an array of n rows and L columns.

$$\begin{array}{cccc} \mathbf{x}_{11} & \mathbf{x}_{21} & \cdots & \mathbf{x}_{L1} \\ \vdots & \vdots & & \vdots \\ \mathbf{x}_{1n} & \mathbf{x}_{2n} & \cdots & \mathbf{x}_{Ln} \end{array}$$

Since each label is the same as one of the different labels $\mathbf{x}_1, \mathbf{x}_2, \ldots$, $\mathbf{x}_L$, this is all possible labels of the unit circle with n numbers. We began this solution by showing that there are exactly $n!$ such labels. Moreover, the array has n rows and L columns, so there are nL labels in this array. Thus,

$$n! = nL$$

so that $L = n!/n = (n-1)!$.

EXAMPLE 14.2.4 There are $(n-1)!$ ways to seat n people around a round table.

14.3 Bracelets

In the above section we counted the number of ways to reserve n seats at a round table. Two sittings at this table are the same if we can rotate one into the other. In this section we consider the number of ways to make a bracelet with different colored beads. This is different from the table problem because, in addition to rotations of the bracelet, you can also flip the bracelet over to make the same bracelet. Thus, in defining what we mean by the same bracelet, we must account for flipping the bracelet as well as rotating it.

Construct a *bracelet* by adding colored beads to a piece of string. The beads are uniformly distributed. This is what you gave/received from your high school sweetheart. It may have had charms dangling from it instead of colored beads, but it was still a bracelet as we described a bracelet. We will efficiently represent bracelets as sequences of the vertices as we have in the diagram below.

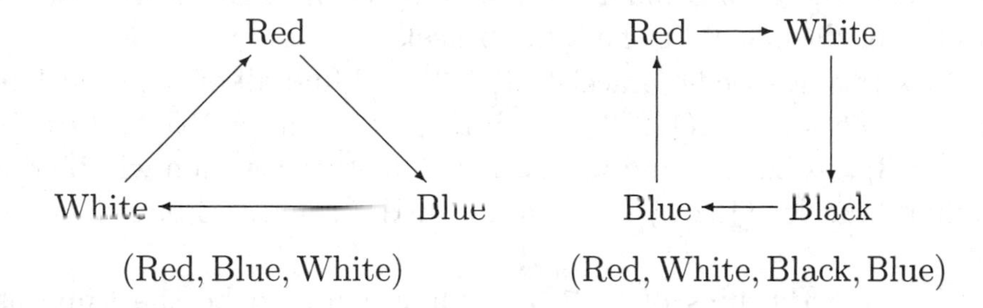

(Red, Blue, White) (Red, White, Black, Blue)

Notice that the sequences follow the paths indicated by the arrows in the bracelets. We feel it is obvious that 2 triangle bracelets A and B are the same if by flipping and rotating A, or just by rotating A, we have produced the three colors on B in the same order that they appear in B.

Let us abstract this notion. A *bracelet made from n different colors* is a label $(x_1, \ldots, x_n)$ of the unit circle with the numbers $\{1, \ldots, n\}$. (Yes, we are using numbers as colors now.) Thus, $(1, 2, 3)$ and $(2, 4, 1, 3)$ are bracelets. The first has 3 different colors, and the second has 4 different colors. The bracelet $(1, 2, 3)$ colors bead 1 with 1, colors bead 2 with 2, and colors bead 3 with 3. The bracelet $(2, 4, 1, 3)$ colors bead 1 with 2, colors bead 2 with 4, colors bead 3 with 1, and colors bead 4 with 3.

Two bracelets B and C made from n different colors are the *same* if we can make C by rotating or flipping B. If bracelets are not the same then they are called *different*. Notice that our definitions of *same* and *different* here apply to bracelets and not to labels. Our different uses of the term *same* will be clear from the language we use. We will refer to *the same bracelets* and not to *the same labels*. This will become as natural as using multiply defined words like *regular* or *hits*. The context tells you which meaning to assign the word.

EXAMPLE 14.3.1 How many bracelets with 3 different colors are there?

Solution: There is exactly 1 bracelet with 3 different colors. Here is how that works.

Consider the bracelet with 3 different colored beads. This is a triple (x_1, x_2, x_3) of 3 different numbers, so there are 3! triples. A number of these will be the same bracelet.

Now take a specific bracelet $(1, 2, 3)$ and find all of the bracelets that are the same as $(1, 2, 3)$. From the picture above, the rotations of $(1, 2, 3)$ are $(2, 3, 1)$ and $(3, 1, 2)$. Any further rotation will be the original bracelet $(1, 2, 3)$ so we go no further. These 3 bracelets are the same.

Examine the flips of $(1, 2, 3)$. They will also be the same as $(1, 2, 3)$. From the picture, a flip of the bracelet $(1, 2, 3)$ about bead 1 will produce the bracelet $(1, 3, 2)$. A flip of the bracelet $(1, 2, 3)$ about bead 2 will produce the bracelet $(3, 2, 1)$. A flip of the bracelet $(1, 2, 3)$ about bead 3 will produce the bracelet $(2, 1, 3)$. These are all of the flips possible *on a bracelet of 3 different colors.* Thus, the colored bracelets $(1, 2, 3)$, $(1, 3, 2)$, $(3, 2, 1)$, and $(2, 1, 3)$ are the same.

Hence, the bracelets that are the same as $(1, 2, 3)$ are the rotations $(1, 2, 3), (2, 3, 1), (3, 1, 2)$ and the flips $(1, 3, 2), (3, 2, 1), (2, 1, 3)$ There are 6 bracelets that are the same as $(1, 2, 3)$. Moreover, we opened this solution by showing that there are exactly 6 triples made from 3 different numbers. Thus, all of the bracelets are the same as $(1, 2, 3)$. That is, there is only 1 bracelet with 3 different colors.

Let $n > 3$ be a whole number. The above example points us to a more general principal that can be used to count the number of bracelets with n different colors. The *beads* of the bracelet are the fixed points on the unit circle to which we have assigned a color (number). An *edge* of a bracelet is the space between 2 consecutive beads.

EXAMPLE 14.3.2 Let $n > 3$ be a whole number. There are $\frac{(n-1)!}{2}$ bracelets with n different colors.

Solution: A bracelet with n different colors is a permutation $(x_1, \ldots, x_n)$ of $\{1, \ldots, n\}$. There are $n!$ such permutations. Next, we find how many bracelets are the same as $(1, \ldots, n)$.

Consider the rotations of $\mathbf{x} = (x_1, \ldots, x_n)$. This colors bead 1 with color x_1. We move bead 1 around the bracelet one bead at a time. This produces bracelets

$$
\begin{array}{ll}
\mathbf{x} = (x_1, \ldots, x_n) & \text{which colors bead 1 with color } x_1, \\
(x_n, x_1, \ldots, x_{n-1}) & \text{which colors bead 2 with color } x_1, \\
(x_{n-1}, x_n, x_1, \ldots, x_{n-2}) & \text{which colors bead 3 with color } x_1, \\
\vdots & \\
(x_2, x_3, \ldots, x_n, x_1) & \text{which colors bead n with color } x_1.
\end{array}
$$

These are all of the rotations of $\mathbf{x}$. One more rotation and we would reproduce $\mathbf{x}$. We rotated color x_1 into each of the n beads, so there are n rotations of $\mathbf{x}$.

Next we find the number of flips of $\mathbf{x}$. Call the line defining a flip the *axis for the flip.* The axis for the flip will divide the circle into 2 halves with the same number N of beads on either side of the axis for the flip. This counts $2N$ beads on either side of the axis. There are 2 cases to consider.

1. Suppose that n is an even number and consider a flip of the bracelet.

(a) If the axis passes through bead 1, then bead 1 is not counted in the $2N$ beads. We have then counted $1 + 2N$ beads. Since n is even, and since n is the number of beads, the axis must pass through another bead. Thus, the flip through bead 1 is through 2 beads, so each flip is associated with 2 beads. Thus, there are $n/2$ different axes of flips that pass through a bead.

(b) Otherwise, the axis for the flip does not contain a bead, and it must pass through opposite edges of the bracelet. There are n sides to the bracelet so there are $n/2$ axes through edges for flips.

In total, if n is even, then there are $n = n/2 + n/2$ flips.

2. Suppose that n is an odd number and consider a flip of the bracelet. As above, the number of beads on either side of the axis for the flip is $2N$. The other beads reside on the axis. If the axis contains a bead then we have counted $1+2N$ beads. The axis would not contain another bead since the number of beads n is odd. Thus,

any flip passes through one bead and an edge of the bracelet. There are n such axes so there are n flips if n is odd.

In any case, there are n flips of the bracelet, so $\mathbf{x}$ is the same as n bracelets found through the application of flips. Putting things together, we use rotations to find n bracelets that are the same as $\mathbf{x}$, and the application of flips gives us n more bracelets that are the same as $\mathbf{x}$. Therefore, there are $2n$ bracelets that are the same as $\mathbf{x}$.

Now, let B be the number of different bracelets made from n different colors, and list the different bracelets.

$$\mathbf{x}_1, \mathbf{x}_2, \ldots, \mathbf{x}_B$$

If $\mathbf{x}_i$ is one of these bracelets then our work in the previous paragraph shows that there are $2n$ bracelets that are the same as $\mathbf{x}_i$. Make a column of the $2n$ bracelets that are the same as $\mathbf{x}_i$.

$$\begin{matrix} \mathbf{x}_{i1} \\ \vdots \\ \mathbf{x}_{i,2n} \end{matrix}$$

Use these columns to make an array of the same bracelets.

$$\begin{matrix} \mathbf{x}_{11} & \mathbf{x}_{21} & \cdots & \mathbf{x}_{B1} \\ \mathbf{x}_{12} & \mathbf{x}_{22} & \cdots & \mathbf{x}_{B2} \\ \vdots & \vdots & & \vdots \\ \mathbf{x}_{1,2n} & \mathbf{x}_{2,2n} & \cdots & \mathbf{x}_{B,2n} \end{matrix}$$

Column 1 is all of the bracelets that are the same as $\mathbf{x}_1$, Column 2 is all of the bracelets that are the same as $\mathbf{x}_2$, and so on. Column B is all of the bracelets that are the same as $\mathbf{x}_B$.

The array has $2n$ rows and B columns so it contains exactly $2nB$ labels. Moreover, each label is the same as some bracelet, and at the beginning of our solution we showed that there are exactly $n!$ labels. Thus, there are $n!$ labels in the array. Hence

$$n! = 2nB,$$

so that there are $B = \dfrac{n!}{2n} = \dfrac{n(n-1)!}{2n} = \dfrac{(n-1)!}{2}$ different bracelets made from n different colors.

14.4 Exercises

The indistinguishable elements referred to in these exercises are typed on a large page. There are enough of them to do these exercises.

1. There is a page of indistinguishable n A's and m B's. Let $n \geq k \geq 0$ $m \geq \ell \geq 0$. In how many ways can you choose exactly k A's and ℓ B's? Answer: 1

2. There is a page of indistinguishable n A's. Let $n \geq k \geq 0$. In how many ways can you choose at least k A? Answer: $n - (k + 1)$

3. There is a page of indistinguishable n A's and m B's. In how many ways can you choose at least 1 A and at least 1 B? Answer: $(n - 1)(m - 1)$

4. Argue without using the formula that there are 3 different bracelets with 4 different colors.

5. Argue without using the formula that there are 12 different bracelets with 5 different colors.

Chapter 15

Algebra and Counting

In this chapter we will investigate algebraic applications of the combinatorics that we have studied to date. This will include the binomial theorem, Pascal's triangle, and some formulas for combination numbers.

15.1 The Binomial Theorem

We will show that for a given whole number $n > 0$, there is a formula for $(1+x)^n$ that includes combination numbers $\binom{n}{k}$ for whole numbers $0 \leq k \leq n$. The symbol $\sum_{k=0}^{k} a_k$ means *add up the numbers* a_k *from* $k = 0$ *to* $k = n$. In more symbols

$$\sum_{k=0}^{k} a_k = a_0 + a_1 + \cdots + a_n.$$

THE BINOMIAL THEOREM 15.1.1 *Let* $n > 0$ *be a whole number. Then*

$$(1+x)^n = \sum_{k=0}^{n} \binom{n}{k} x^k.$$

In other words,

$$(1+x)^n = \binom{n}{0} x^0 + \binom{n}{1} x^1 + \binom{n}{2} x^2 + \cdots + \binom{n}{n} x^n$$

$$= 1 + nx + \binom{n}{2} x^2 + \binom{n}{3} x^3 + \cdots + x^n.$$

The term x comes from the fact that $\binom{n}{1} = n$. The term x^n comes from the fact that $\binom{n}{n} = 1$. For instance,

$$\begin{aligned} (1+x)^2 &= \binom{2}{0} + \binom{2}{1} x^1 + \binom{2}{2} x^2 \\ &= 1 + 2x + x^2 \end{aligned}$$

and similarly

$$\begin{aligned} (1+x)^3 &= \binom{3}{0} + \binom{3}{1} x^1 + \binom{3}{2} x^2 + \binom{3}{3} x^3 \\ &= 1 + 3x + 3x^2 + x^3. \end{aligned}$$

So let us examine where this formula comes from.

EXAMPLE 15.1.2 Prove the Binomial Theorem.

Solution: Let $n > 0$ be a whole number. Write the power $(1+x)^n$ as a product of $1 + x$'s.

$$(1+x)^n = \underbrace{(1+x)\cdots(1+x)}_{n} \tag{15.1}$$

When written as a sum, $(1+x)^n$ becomes a polynomial

$$(1+x)^n = a_0 + a_1 x + a_2 x^2 + \cdots + a_n x^n \tag{15.2}$$

for some real numbers $a_0, a_1, \ldots, a_n$. Let us find these coefficients using combination numbers.

In a generalized foil, the terms in the sum for $(1+x)^n$ are formed by adding the products of n factors $d_1 \cdots d_n$ where each d_i is chosen as a symbol in the i-th copy of $(1+x)$ in (15.1). For example, in $(1+x)^3 = (1+x)(1+x)(1+x)$ we can form $1 \cdot 1 \cdot 1$, which is formed by taking 1 from the first, second, and third copy of 1 in $(1+x)(1+x)(1+x)$. We can form $1 \cdot x \cdot 1$, which is formed by taking

1 form the first and third copy of $(1+x)$ in $(1+x)(1+x)(1+x)$, and choosing x form the second copy of $(1+x)$ in $(1+x)(1+x)(1+x)$.

The number of occurrences of x in the product $d_1 \cdots d_n$ determines which power of x the product $d_1 \cdots d_n$ will contribute to. For example, the following products contribute to the a_2x^2 term in the sum $(1+x)^3 = a_0 + a_1x + a_2x^2 + a_3x^3$.

$$x \cdot x \cdot 1 \qquad x \cdot 1 \cdot x \qquad 1 \cdot x \cdot x$$

This is all of the words $d_1d_2d_3$ that contribute to a_2x^2. When added together they add to the term a_2x^2 as follows.

$$a_2x^2 = x \cdot x \cdot 1 + x \cdot 1 \cdot x + 1 \cdot x \cdot x = x^2 + x^2 + x^2 = 3x^2.$$

We will try to apply this intuition to finding the terms a_kx^k in (15.2).

Let $0 \leq k \leq n$ and find the a_kx^k term of $(1+x)^n$ in (15.2). Suppose that the product $d_1 \cdots d_n$ contributes to a_kx^k. Then $d_1 \cdots d_n$ as a word from $\{1, x\}$ contains exactly k x's, and exactly $n-k$ 1's. We count the number of words $d_1 \cdots d_n$ by choosing the k places for x in $d_1 \cdots d_n$. This is done in $\binom{n}{k}$ ways. Every possible word $d_1 \cdots d_n$ that contains exactly k x's is constructed in this way. Thus

$$a_kx^k = \binom{n}{k} x^k.$$

Here are some concrete examples of how we just calculated a_kx^k.

1. To find a_0 we look for all of the words $d_1 \cdots d_n$ that do not have x's, and that have n 1's. There is only 1 such word, $\underbrace{1 \cdots 1}_{n}$, so $a_0 = 1$.

2. We find a_1x. The words $d_1 \cdots d_n$ that contribute to a_1x have exactly 1 x and $n-1$ 1's. Choose the 1 place in $d_1 \cdots d_n$ to contain the x in $\binom{n}{1}$ ways. The sum of these words is a_1x so

$$a_1x = \binom{n}{1} x = nx.$$

3. Find a_2x^2. The words $d_1 \cdots d_n$ that contribute to a_2x^2 have exactly 2 x's and $n-2$ 1's. Choose the 2 places in $d_1 \cdots d_n$ to contain the x in $\binom{n}{2}$ ways. The sum of these words is a_2x^2 so that

$$a_2x^2 = \binom{n}{2} x^2.$$

To conclude our proof of the Binomial Theorem we observe that by (15.2) and our values for a_kx^k we have found that

$$(1+x)^n = \sum_{k=0}^{n} \binom{n}{k} x^k =$$

$$\binom{n}{0} + \binom{n}{1} x + \cdots + \binom{n}{k} x^k + \cdots + \binom{n}{n} x^n$$

This is what the Binomial Theorem requires, so the proof is complete.

15.2 Identities

As an example of how the Binomial Theorem can be used, we offer several connection between $(1+x)^n$ and combination numbers.

EXAMPLE 15.2.1 Let $n > 0$ be a whole number. Then

$$2^n = \binom{n}{0} + \binom{n}{1} + \cdots + \binom{n}{k} + \cdots + \binom{n}{n}.$$

Solution: Let $x = 1$ in the Binomial Theorem. Then

$$\begin{aligned} 2^n &= (1+1)^n \\ &= \binom{n}{0} + \binom{n}{1} 1 + \cdots + \binom{n}{k} 1^k + \cdots + \binom{n}{n} 1^n \\ &= \binom{n}{0} + \binom{n}{1} + \cdots + \binom{n}{k} + \cdots + \binom{n}{n}. \end{aligned}$$

This is what we had to show.

Every student of Algebra 3 has seen *Pascal's triangle.* A seventeenth century mathematician named B. Pascal showed that the coefficients of $(1+x)^n$ can be found in the n-th row of the (endless) triangle

$$\begin{array}{ccccccccccc} & & & & 1 & & 1 & & & & \\ & & & 1 & & 2 & & 1 & & & \\ & & 1 & & 3 & & 3 & & 1 & & \\ & 1 & & 4 & & 6 & & 4 & & 1 & \\ 1 & & 5 & & 10 & & 10 & & 5 & & 1 \end{array}$$

The rows of this triangle continue indefinitely. The key to producing the entries is that a number c in the array is the sum of the numbers above c. Thus 6 is the sum of 3 and 3 above it, and 3 is the sum of the numbers 1 and 2 above 3. To produce the next row, row number 6, simply add entries in row 5 as follows:

$$\begin{array}{ccccccccccccc} & 1 & & 5 & & 10 & & 10 & & 5 & & 1 & \\ 1 & & 6 & & 15 & & 20 & & 15 & & 6 & & 1 \end{array}$$

The purpose behind this triangle is that row n contains the coefficients of $(1+x)^n$ in the order in which they appear in $(1+x)^n$. The first row is

$$1 \quad 1$$

and

$$(1+x)^1 = 1 + 1x.$$

The 1's in the first row of Pascal's triangle are found as coefficients of $(1+x)^1$.

Row 2 of Pascal's triangle is

$$1 \quad 2 \quad 1$$

and

$$(1+x)^2 = 1 + 2x + x^2.$$

The entries in row 2 of Pascal's triangle are the coefficients of $(1+x)^2$.

The same holds for any row. Thus the coefficients of $(1+x)^5$ are given in row 5 of Pascal's triangle. Hence,

$$(1+x)^5 = 1 + 5x + 10x^2 + 10x^3 + 5x^4 + x^5$$

where the numbers $1, 5, 10, 10, 5, 1$ come from row 5 of the triangle given above.

There are some simple formulas given by Pascal's triangle. The first and last entries of a row are 1. These are the coefficients of 1 and x^n in $(1+x)^n$. Since the Binomial Theorem shows us that $\binom{n}{0}$ and $\binom{n}{n}$ are the coefficients of 1 and x^n in $(1+x)^n$, we find that

$$\binom{n}{0} = 1 \text{ and } \binom{n}{n} = 1.$$

These can be verified by hand since $\binom{n}{0}$ and $\binom{n}{n}$ are simple fractions to calculate.

The second and second to the last entry in row n of Pascal's triangle is n. Then n is the coefficient of x and x^{n-1} in $(1+x)^n$. By the Binomial Theorem the coefficient of x is $\binom{n}{1}$ and the coefficient of x^{n-1} is $\binom{n}{n-1}$. Thus

$$\binom{n}{1} = \binom{n}{n-1} = n.$$

This identity can also be found by counting the number of ways to choose 1 element from a set of n elements, and by counting the number of ways to choose 1 element to be left behind when choosing $n-1$ elements from n.

EXAMPLE 15.2.2 Let $0 \leq k \leq n$ be whole numbers. Show that

$$\binom{n}{k} + \binom{n}{k+1} = \binom{n+1}{k+1}.$$

Solution: By the Binomial Theorem the coefficient of x^k in $(1+x)^n$ is $\binom{n}{k}$, and we know that the coefficients of x^k in $(1+x)^n$ are in the n-th row of Pascal's triangle. The relevant portion of rows n and $n+1$ of Pascal's triangle are given below.

$$\begin{array}{ccccccc} \ldots & & \binom{n}{k} & & \binom{n}{k+1} & & \ldots \\ \ldots\binom{n+1}{k} & & & \binom{n+1}{k+1} & & & \binom{n+1}{k+2}\ldots \end{array}$$

The way we generate entries in Pascal's triangle is to add two adjacent entries to get the entry below them. Thus

$$\binom{n}{k} + \binom{n}{k+1} = \binom{n+1}{k+1}.$$

This is the formula we were asked to justify.

EXAMPLE 15.2.3 Let $0 \leq k \leq n$ be whole numbers. Show that

$$\binom{n}{k} + 2\binom{n}{k+1} + \binom{n}{k+2} = \binom{n+2}{k+2}.$$

Solution: Using Example 15.2.2 we can write

$$\binom{n}{k} + \binom{n}{k+1} = \binom{n+1}{k+1}$$

and

$$\binom{n}{k+1} + \binom{n}{k+2} = \binom{n+1}{k+2}.$$

Another application of Example 15.2.2 yields

$$\binom{n+1}{k+1} + \binom{n+1}{k+2} = \binom{n+2}{k+2}.$$

Then $\binom{n}{k} + 2\binom{n}{k+1} + \binom{n}{k+2}$

$$= \left(\binom{n}{k} + \binom{n}{k+1}\right) + \left(\binom{n}{k+1} + \binom{n}{k+2}\right)$$

$$
\begin{aligned}
&= \binom{n+1}{k+1} + \binom{n+1}{k+2} \\
&= \binom{n+2}{k+2}.
\end{aligned}
$$

This is what we set out to show.

EXAMPLE 15.2.4 Let $0 \le k \le n$ be fixed whole numbers. Show that

$$
\binom{k}{k} + \binom{k+1}{k} + \binom{k+2}{k} + \cdots + \binom{n}{k} = \binom{n+1}{k+1}.
$$

Solution: By Example 15.2.2, for each whole number $0 \le m \le k$, we have

$$
\binom{m}{k} + \binom{m}{k+1} = \binom{m+1}{k+1}
$$

or more to the point

$$
\binom{m}{k} = \binom{m+1}{k+1} - \binom{m}{k+1}. \tag{15.3}
$$

Then

$$
\begin{aligned}
&\binom{k}{k} + \binom{k+1}{k} + \binom{k+2}{k} + \cdots + \binom{n}{k} \\
&= 1 + \left(\binom{k+2}{k+1} - \binom{k+1}{k+1}\right) + \\
&\qquad \left(\binom{k+3}{k+1} - \binom{k+2}{k+1}\right) \\
&\qquad\qquad + \cdots + \left(\binom{n+1}{k+1} - \binom{n}{k+1}\right).
\end{aligned}
$$

This is a telescoping sum like one you would encounter in a first year calculus course, had you taken that course in the twentieth century. The cancellation of terms begin with

$$
1 - \binom{k+1}{k+1} = 0
$$

and then for each $0 \le m \le n$ there are terms

$$\binom{m-1}{k} \text{ and } \binom{m}{k}$$

in our first sum. By (15.3) these terms expand into

$$\binom{m}{k+1} - \binom{m-1}{k+1} \text{ and } \binom{m+1}{k+1} - \binom{m}{k+1}$$

in the second sum. Thus the sum $\binom{m-1}{k} + \binom{m}{k}$ introduces the cancellation

$$\begin{aligned} & \left(\binom{m}{k+1} - \binom{m-1}{k+1}\right) + \left(\binom{m+1}{k+1} - \binom{m}{k+1}\right) \\ = & -\binom{m-1}{k+1} + \binom{m+1}{k+1} \end{aligned}$$

in the second sum. Hence, because $0 \le m \le n$, we have cancellation of every symbol $\binom{m}{k+1}$ except for $\binom{n+1}{k+1}$. That is,

$$\binom{k}{k} + \binom{k+1}{k} + \binom{k+2}{k} + \cdots + \binom{n}{k} = \binom{n+1}{k+1}.$$

EXAMPLE 15.2.5 Verify that

$$\binom{1}{1} + \binom{2}{1} + \binom{3}{1} = \binom{4}{2}$$

by calculating each of the combination numbers.

EXAMPLE 15.2.6 By the formula in Example 15.2.2, we have

$$\binom{1}{1} + \binom{2}{1} + \cdots + \binom{n}{1} = \binom{n+1}{2}$$

which is the familiar Gaussian sum

$$1 + 2 + \cdots + n = \frac{(n+1)n}{2}.$$

15.3 Exercises

1. Using Pascal's triangle show that $\binom{n}{k} = \binom{n}{n-k}$.

2. Use Example 15.2.1 to show that 2^n is the number of subsets of a set with exactly n elements.

3. Verify that $3^n = 1 + 2n + 2^2\binom{n}{2} + \cdots + 2^n$ for each whole number $n > 0$.

Chapter 16

Derangements

Let us consider the following counting problem. Let $n > 1$ be a whole number. There are n men at a house party. They pile their coats on the house bed. At some point a bell goes off and the men leave the house in such a hurry that they grab their coats at random, not considering which coat is theirs.

QUESTION: In how many ways can each man get someone else's coat?

For example, suppose there are 3 men and three coats. We will name the men $1, 2$, and 3. The coats will be places in a triple $(\, , \, , \,)$. So the first place in the triple represents the first coat, the second place represents the second coat, and the third place represents the third coat. The symbol $(2, 3, 1)$ means that man 2 gets coat 1, man 3 gets coat 3, and man 1 gets coat 3. In this way, the assignments of which man gets which coat is a permutation of the numbers $1, 2$, and 3.

By saying that each man should get someone else's coat, we disallow any permutation in which the man m sits in the place representing coat m. So 1 does not sit in place 1, 2 does not sit in place 2, and 3 does not sit in place 3. A permutation with this property, that no number k sits in the k-th place, is called a *derangement.*

The complement of a derangement is a permutation that has at least 1 number m that sits in place m. For instance, $(3, 2, 1)$ fixes the number 2 since 2 sits in place 2. We would say that $(3,2,1)$ has *at least 1 fixed point.* The permutation $(1,2,3)$ fixes the numbers 1, 2, and 3 so $(1,2,3)$ has at least 1 fixed pint.

In this chapter we will answer the question above by counting the number of permutations that fix at least one number in $\{1,\ldots,n\}$, and then counting the number of derangements of the set $\{1,\ldots,n\}$.

16.1 Mathematical Induction

We begin with the counting method that we will use throughout this chapter. It is called *Mathematical Induction.* Here is how it works.

MATHEMATICAL INDUCTION 16.1.1 Let $f(n)$ be a function, or specifically a formula, that is defined for every integer $n \geq 0$. Then to prove that $f(n)$ is True for every $n \geq 0$ we proceed as follows.

1. *Initial Step:* Show that $f(0)$ or True, that $f(n_0)$ is True for some fixed integer n_0.

2. *Induction Hypothesis:* Assume that $f(k)$ is True for some integer $k \geq 0$. This itself is a True statement since by the *Initial Step*, $f(n_0)$ is True for some integer n_0.

3. *Induction Proof:* Using the Induction Hypothesis, show that $f(k+1)$ is True.

Once these three steps have been established, we conclude with

By Mathematical Induction, $f(n)$ is True for every integer $n \geq 0$.

Notice what this does. In order to prove that the infinitely many statements $f(0), f(1), f(2), f(3), \ldots$ are True, we need only establish the Truth of the two simple statements the *Initial Step* and the *Induction Proof.* Here are a few examples.

EXAMPLE 16.1.2 1. To show that $2n+1 < 3n+2$ for al integers $n \geq 0$, proceed as follows.

Initial Step: $f(0)$ is the inequality $2 \cdot 0 + 1 < 3 \cdot 0 + 2$, which is True. Thus the Initial Step is proved.

Induction Hypothesis: Suppose that $2k + 1 < 3k + 2$ for some integer $k \geq 0$.

Induction Proof: Add $(k + 1) + 1$ to $2k + 1$ to produce the inequality

$$2(k + 1) + 1 < 2(k + 1) + 1 + (k + 1) + 1 = 3(k + 1) + 2.$$

Then by Mathematical Induction, $2n+1 < 3n+2$ for all integers $n \geq 0$.

2. Given integers $n \geq k \geq 0$, the statement $f(n)$ is

$$\frac{n!}{k!} = n(n - 1) \cdots (n - k + 1).$$

We will use Mathematical Induction to prove that given an integer $k \geq 0$, *$f(n)$ is True for each integer $n \geq k$.*

Fix an integer $k \geq 0$.

Initial Step: Given $n = k$ we see that

$$\frac{n!}{k!} = \frac{k!}{k!} = 1 = k \cdots (k - k + 1) = k \cdots (n - k + 1).$$

This proves the Initial Condition.

Induction Hypothesis: Assume that there is an integer $m \geq k$ such that $f(m)$ is True: $\dfrac{m!}{k!} = n \cdots (n - k + 1)$.

Induction Proof: We must show that $\dfrac{(m + 1)!}{k!} = (m+1) \cdots ((m+1) - k + 1)$. By the Induction hypothesis, $\dfrac{m!}{k!} = m \cdots (m - k + 1)$. Multiply both sides by $m + 1$ to find that

$$\begin{aligned} (m + 1) \cdot \frac{m!}{k!} &= (m + 1) \cdot m \cdots (m - k + 1) \\ \frac{(m + 1) \cdot m!}{k!} &= (m + 1)m \cdots ((m + 1) - k + 1) \\ \frac{(m + 1)!}{k!} &= (m + 1) \cdots ((m + 1) - k + 1). \end{aligned}$$

This is the statement $f(m+1)$, so by Mathematical Induction, we have proved that $\frac{n!}{k!} = n \cdots (n-k+1)$ for all integers $n \geq 0$. This completes the proof.

There is another form of Mathematical Induction that goes as follows. Notice that the Induction Hypothesis below assumes that $f(k)$ is true for some integer $k \geq 0$. This is different from the *Induction Hypothesis* given in Mathematical Induction 16.1.1 where we assume that for some fixed integer $n \geq 0$, $f(k)$ is True for all integers $n > k$.

MATHEMATICAL INDUCTION 16.1.3 Let $f(n)$ be a function, or specifically a formula, that is defined for every integer $n \geq 0$. Then to prove that $f(n)$ is True for every $n \geq 0$ we proceed as follows.

1. *Initial Step:* Show that $f(0)$ is True.
2. *Induction Hypothesis:* Let $n \geq 0$ be an integer. Assume that $f(k)$ is True for each integer $n > k$.
3. *Induction Proof:* Using the Induction Hypothesis, show that $f(k+1)$ is True.

Once these three steps have been established, we conclude with

By Mathematical Induction, $f(n)$ is True for every integer $n \geq 0$.

The examples we offer are classic results about numbers.

A *prime number* is an integer $p \geq 2$ that is evenly divisible only by 1 and itself. Notice that prime numbers are larger than 2, and thus that *1 is not a prime number.* The reason for this is that in a field, $0 \neq 1$. We say that an integer $n \geq 2$ *possesses a prime divisor* if there is a prime number p such that $n = pc$ for some integer $c \geq 1$.

The next theorem proves what everyone out of the sixth grade should know, that every number possesses a prime divisor.

THEOREM 16.1.4 *Let $n \geq 2$ be an integer. Then n possesses a prime divisor.*

Proof: *Initial Step:* The number $n = 2$ is divisible by 2.

Induction Hypothesis: Let $n \geq 0$ be an integer. Assume for each integer $n > k$, that k divisible by a prime number.

Induction Proof: Consider the prime divisors of n. If n is a prime number then we are done as n is divisible by the prime n. Otherwise, $n = ab$ for some integers $a, b \geq 2$. Then $a = n/b < n$ are integers, so by the Induction Hypothesis, a is divisible by a prime number p. Thus $a = pc$ for some integer c, whence $n = ab = pcb$ is divisible by the prime p, and this completes the Induction Proof.

Hence, by Mathematical Induction, each integer $n \geq 2$ is divisible by a prime number. This completes the proof.

The next theorem is fundamental to all of arithmetic. It shows us that prime factorization is unique for every positive integer ≥ 2. While there may be other proofs of this theorem, we choose the instructive one that employs Mathematical Induction.

A *prime factorization of* n is a product $n = p_1 \cdots p_t$ in which $p_1 \leq \ldots \leq p_t$ are prime numbers. The product is said to be *unique* if $p_1 \leq \ldots \leq p_t$ is the list of primes in every prime factorization of n.

The FUNDAMENTAL THEOREM of ARITHMETIC 16.1.5 *If $n \geq 2$ is an integer then n possesses a unique prime factorization.*

Proof: *Initial Step:* Let $n = 2$. Then n is its own unique prime factorization, 2. This proves the Initial Step.

Induction Hypothesis: Assume that we have found an integer $n \geq 2$ such that each integer $n > k \geq 2$ possesses a unique prime factorization.

Induction Proof: Assume the Induction Hypothesis. If n is a prime number, then as in the Initial Step, n is its own unique prime factorization. So assume that n is not a prime. By Theorem 16.1.4, there is a prime number p and an integer $k \geq 2$ such that $n = pk$. Since primes are at least 2, $k < n$. The Induction Hypothesis then states that k possesses a unique prime factorization $p_1 \cdots p_t$ for some finite list of primes $p_1 \leq \cdots \leq p_t$.

If q is another prime factor of n then because p is prime, either $p = q$ or q divides k. Because the prime factorization of k is unique, $q = p_i$ for some $i = 1, \ldots, t$. Hence, in any prime factorization of n the primes that occur are on the list $p, p_1, \ldots, p_t$, and therefore the prime factorization $n = p \cdot p_1 \cdots p_t$ is unique. This concludes the Induction Proof.

Then by Mathematical Induction, each integer $n \geq 2$ possesses a unique prime factorization. This completes the proof.

16.2 Fixed-Point Theorems

A *permutation* is an arrangement of the sequence $(1, 2, \ldots, n)$ of the numbers $\{1, \ldots, n\}$. Thus, $(2, 1)$ is a permutation of $\{1, 2\}$ and $(1, 3, 4, 2)$ is a permutation of $\{1, 2, 3, 4\}$. Some permutations stand out. We will say that a permutation fixes k if k is in the k-th place. For instance, the permutation $(1, 2)$ fixes both 1 and 2 since 1 is in place number 1 and 2 is in place number 2. The permutation $(2, 3, 1, 4)$ fixes 4 and no other number, as we can see by reading the permutation.

Let $1 \leq k \leq n$ be a whole number. We say that a permutation $(x_1, \ldots, x_n)$ of $\{1, \ldots, n\}$ *fixes* k if $x_k = k$. A permutation with $x_2 = 2$ fixes 2, and a permutation such that $x_{10} = 10$ fixes 10. We say that the permutation has *at least 1 fixed point* if the permutation fixes k for some $1 \leq k \leq n$. The permutation is a *derangement* if it has no fixed points.

The problem we will solve is to count the number of permutations with at least 1 fixed point, and the number of permutations without fixed points. Let

$F_n =$ the number of permutations of $\{1, \ldots, n\}$
with at least 1 fixed point.

We must find F_2, F_3, and in general F_n for any whole number $n > 1$.

EXAMPLE 16.2.1 Let $n > 1$ be a whole number. We count the number of derangements of $\{1, \ldots, n\}$.

Solution: Let $\mathcal{U}$ be the set of all permutations of $\{1, \ldots, n\}$, and let FP_n be the set of permutations of $\{1, \ldots, n\}$ that have at least 1 fixed point. In the by now familiar method, $n(\mathcal{U}) = n!$, and by our notation

$$n(FP_n) = F_n.$$

The complement of FP_n is the set FP'_n of all permutations of $\{1, \ldots, n\}$ that have no fixed points. Then by the complement formula

$$n(FP'_n) = n(\mathcal{U}) - n(FP_n) = n! - F_n.$$

In other words,

$$n! - F_n = \text{The number of derangements of } \{1, \ldots, n\}.$$

There is a formula for F_n but it relies on a knowledge of the smaller values $F_{n-1}, \ldots, F_2$. This is called a *recursive formula*, and it is common in the mathematics associated with computer science. We begin by finding F_2.

EXAMPLE 16.2.2 Find F_2.

Solution: Write out the 2 permutations of $\{1, 2\}$.

$$(1, 2), (2, 1)$$

The first fixes 1 and 2, and $(2, 1)$ is without fixed points since 2 is in place 1 and 1 is in place 2. Thus,

$$F_2 = 1.$$

There is exactly 1 permutation with at least 1 fixed point.

That is a start. This F_2 will be the stopping point in the descent as we find F_n. More details follow.

EXAMPLE 16.2.3 Find F_3 = the number of permutations of $\{1, 2, 3\}$ that fix at least 1 point.

Solution: A permutation of $\{1, 2, 3\}$ is a triple (a, b, c) with $a, b, c \in \{1, 2, 3\}$. Choose a permutation. We work in cases.

1. There is only 1 permutation that fixes all numbers, $(1, 2, 3)$. If the chosen permutation fixes 2 numbers then it must also fix the last number. Thus, the only permutation of $\{1, 2, 3\}$ that fixes at least 2 numbers is $(1, 2, 3)$.

2. Suppose the chosen permutation fixes exactly 1 number. There are $\binom{3}{1} = 3$ choices for the number fixed. Since there will be exactly 1 fixed number, the remaining 2 places have no fixed points. We count this by counting the number of permutations of 2 numbers that are without a fixed point. By Examples 16.2.1 and 16.2.2 that number is

$$2! - F_2 = 1.$$

Hence, by adding the results for cases 1 and 2, there are

$$\boxed{F_3 = \binom{3}{1} \cdot (2! - F_2) + 1 = 4}$$

permutations of $\{1, 2, 3\}$ with at least 1 fixed point. The summand 1 corresponds to the permutation of $\{1, 2, 3\}$ that fixes at least 2 places.

Now we make the our mathematical move to the most general case.

EXAMPLE 16.2.4 Let $n > 1$ be a whole number. We find

$$F_n = \text{the number of permutations of } \{1, \ldots, n\} \text{ with at least 1 fixed point.}$$

Solution: To count F_n we begin by noting that a permutation with at least 1 fixed point then has exactly k fixed points for some $1 \leq k \leq n$. To construct a permutation that fixes exactly k numbers, choose k numbers to be fixed in $\binom{n}{k}$ ways. In the remaining

$n-k$ places we must insert a derangement of the remaining $n-k$ numbers. By Example 16.2.1,

$$(n-k)! - F_{n-k} = \text{the number of derangements of } \{1, \ldots, n-k\}.$$

Hence, the multiplication principal shows us that the number of permutations of $\{1, \ldots, n\}$ that have exactly k fixed point is

$$\binom{n}{k}((n-k)! - F_{n-k}). \tag{16.1}$$

In our context, the predicates *exactly* $n-1$ *fixed points* and *exactly* n *fixed points* define the same singleton set $\{(1, \ldots, n)\}$. Furthermore, from our extensive work in counting permutations, we know that the predicate *at least 1 fixed point* can be broken into the smaller cases

exactly 1 fixed point
exactly 2 fixed points
⋮
exactly $n-2$ fixed points
exactly $n-1$ or n fixed points.

By formula (16.1), we have

$$\begin{aligned}
n(\text{exactly 1 fixed point}) &= \binom{n}{1}((n-1)! - F_{n-1}) \\
n(\text{exactly 2 fixed points}) &= \binom{n}{2}((n-2)! - F_{n-2}) \\
&\vdots \\
n(\text{exactly } n-2 \text{ fixed points}) &= \binom{n}{n-2}((2)! - F_2) \\
n(\text{exactly } n-1 \text{ or } n \text{ fixed points}) &= 1
\end{aligned}$$

and in general we have

$$\begin{aligned}
&n(\text{at least 1 fixed point}) = \\
&\quad n(\text{exactly 1 fixed point}) + \cdots + n(\text{exactly } n-2 \text{ fixed point}) + 1.
\end{aligned}$$

Therefore, the number of permutations of $\{1, \ldots, n\}$ with at least 1 fixed point is

$$F_n = \binom{n}{1}((n-1)! - F_{n-1}) + \binom{n}{2}((n-2)! - F_{n-2}) + \cdots + \binom{n}{n-2}((2)! - F_2) + 1$$

Since we know the number F_n of permutations that fix at least 1 element in $\{1, \ldots, n\}$, we can take its complement and find the number of permutations that do not fix any element of $\{1, \ldots, n\}$. This is, of course, the number of derangements of the set $\{1, \ldots, n\}$.

EXAMPLE 16.2.5 Let $n > 1$ be a whole number.

The number of derangements of $\{1, \ldots, n\}$ is

$$\begin{aligned} n! - F_n &= n! + \binom{n}{1}(F_{n-1} - (n-1)!) \\ &\quad + \binom{n}{2}(F_{n-2} - (n-2)!) \\ &\quad + \cdots + \binom{n}{n-2}(F_2 - (2)!) - 1 \end{aligned}$$

Let us use this formula to find F_n for some small values of n.

EXAMPLE 16.2.6 Find F_4.

Solution: In Examples 16.2.2 and 16.2.3 we found that $F_2 = 1$ and $F_3 = 4$. Hence, by our formula for F_n we have

$$\begin{aligned} F_4 &= \binom{4}{1}((3)! - F_3) + \binom{4}{2}((2)! - F_2) + 1 \\ &= 4(2) + 6(1) + 1 = 15 \end{aligned}$$

EXAMPLE 16.2.7 Find the number of derangements of $\{1, 2, 3, 4\}$.

Solution: That number is

$$4! - F_4 = 24 - 15 = 9.$$

EXAMPLE 16.2.8 Find F_5.

Solution: In Examples 16.2.2, 16.2.3, and 16.2.6 we found that $F_2 = 1$, $F_3 = 4$, and $F_4 = 15$. Hence, by our formula for F_n we have

$$\begin{aligned} F_5 &= 5((4)! - F_4) + 10((3)! - F_3) + 20((2)! - F_2) + 1 \\ &= 5(24 - 15) + 10(6 - 4) + 20(2 - 1) + 1 \\ &= 86 \end{aligned}$$

EXAMPLE 16.2.9 Find the number of derangements of $\{1, 2, 3, 4, 5\}$.

Solution: That number is

$$5! - F_5 = 120 - 86 = 34.$$

16.3 His Own Coat

Let us restate the question that began this chapter.

EXAMPLE 16.3.1 Let $n > 1$ be a whole number. There are m men at a house party. They have taken off their coats and piled them 1 the house bed. At some point a bell goes off and they rush to the house bed, picking up a coat without looking at it. In how many ways can the coats be picked up so that each man gets someone else's coat?

Solution: Each man is named with a number from $\{1, \ldots, n\}$. No two men get the same name. Let the places in the n-tuple $(\ , \ldots, \ )$ represent the coats on the bed. So place 1 in the n-tuple represents 1's coat, place k in the n-tuple represents k's coat, and place n in the n-tuple represents n's coat.

If man k gets ℓ's coat then we write k in place ℓ. For example $(4, 3, 2, 1)$ represents the fact that man 4 got 1's coat, man 3 got 2's coat, man 2 got 3's coat, and man 1 got 4's coat. Therefore, if each man gets someone else's coat then we have constructed a derangement of $\{1, \ldots, n\}$. And each derangement represents men that have someone else's coat.

Thus, to solve the problem we need only count the number of derangements of $\{1, \ldots, n\}$. **The number of ways that each**

of n men can take a coat not their own is the number of derangements of $\{1, \ldots, n\}$, which is

The number of ways n men will take someone else's coat is

$$n! - F_n = n! + \binom{n}{1}(F_{n-1} - (n-1)!) + \binom{n}{2}(F_{n-2} - (n-2)!) + \cdots + \binom{n}{n-2}(F_2 - 2!) - 1$$

EXAMPLE 16.3.2 The number of fixed-point permutations of $\{1, 2, 3\}$ is $F_3 = 4$, as we saw in Example 16.2.3. Then the number of derangements of $\{1, 2, 3\}$ is $3! - F_3 = 2$. Thus, there are exactly 2 ways that 3 men can leave the house party with someone else's coat. Those 2 men to coat assignments are $(2, 3, 1)$ and $(3, 1, 2)$.

EXAMPLE 16.3.3 The number of fixed-point permutations of $\{1, 2, 3, 4\}$ is $F_4 = 15$, as we saw in Example 16.2.6. Then the number of derangements of $\{1, 2, 3, 4\}$ is $4! - F_4 = 9$. Thus, there are exactly 9 ways that 4 men can leave the house party with someone else's coat. Try to write down those 9 men to coat assignments. One of them is $(4, 1, 2, 3)$.

16.4 Inclusion/Exclusion for Many Sets

You are familiar with the **inclusion/exclusion formula**

$$(A_1 \cup A_2) = n(A_1) + n(A_2) - n(A_1 \cap A_2) \tag{16.2}$$

for two sets A_1, A_2. Given the n finite sets $A_1, \ldots, A_n$, we can apply the inclusion/exclusion formula for two sets to determine the **inclusion/exclusion formula for n sets**. This will determine the number $n(A_1 \cup \cdots \cup A_n)$ of elements in $A_1 \cup \cdots \cup A_n$. Said formula is boxed in the next theorem.

THEOREM 16.4.1 *Let $A_1, \ldots, A_n$ be n finite sets. Then*

$$n(A_1 \cup \cdots \cup A_n)$$
$$= \sum_{t=1}^{n} (-1)^{t-1} \sum_{1 \le i_1 < \cdots < i_t \le n} n(A_{i_1} \cap \cdots \cap A_{i_t})$$
$$= (-1)^0 \sum_{i=1}^{n} n(A_i) \; + \; (-1)^1 \sum_{1 \le i < j \le n} n(A_i \cap A_j)$$
$$+ \cdots + \; (-1)^{n-1} n(A_1 \cap \cdots \cap A_n).$$

Proof: For notational purposes, given an integer $k \geq 2$, let

$$B_k = A_1 \cup \cdots \cup A_k$$

of k sets $A_1, \ldots, A_k$.

Our proof is by mathematical induction. It begins by observing that the inclusion/exclusion formula (16.2) for 2 sets is the above boxed formula

$$n(A_1 \cup A_2) = \left(\sum_{i=1}^{2} n(A_i) \right) + (-1)^1 n(A_1 \cap A_2).$$

The next step in mathematical induction is to assume that for some integer $k \geq 2$ that *the inclusion/exclusion formula*

$$n(B_k) \;\; = \;\; \sum_{t=1}^{k} (-1)^{t-1} \sum_{1 \le i_1 < \cdots < i_t \le k} n(A_{i_1} \cap \cdots \cap A_{i_t}) \qquad (16.3)$$

for $n(B_k)$ is True for any k sets. After all, by (16.2), this assumption is True of the integer $k = 2$.

We must now establish the Truth of the inclusion/exclusion formula for $k + 1$ sets. Once established, we will have shown that the inclusion/exclusion formula for n sets holds for every integer $n \geq 2$.

Because $B_{k+1} = A_1 \cup \cdots \cup A_{k+1}$, (16.2) shows that

$$n(B_{k+1}) = n(B_k \cup A_{k+1}) = n(B_k) + n(A_{k+1}) - n(B_k \cap A_{k+1}). \qquad (16.4)$$

Notice that

$$\begin{aligned} B_k \cap A_{k+1} &= (A_1 \cup \cdots \cup A_k) \cap A_{k+1} \\ &= (A_1 \cap A_{k+1}) \cup \cdots \cup (A_k \cap A_{k+1}) \end{aligned}$$

is a union of k sets. Then an application of the assumed formula (16.3) yields

$$\begin{aligned} n(B_k \cap A_{k+1}) &= n((A_1 \cap A_{k+1}) \cup \cdots \cup (A_k \cap A_{k+1})) \\ &= \sum_{t=1}^{k} (-1)^{t-1} \sum_{1 \le i_1 < \cdots < i_t \le k} n((A_{i_1} \cap A_{k+1}) \cap \cdots \cap (A_{i_t} \cap A_{k+1})) \\ &= \sum_{t=1}^{k+1} (-1)^{t-1} \sum_{1 \le i_1 < \cdots < i_{t-1} \le k} n(A_{i_1} \cap \cdots \cap A_{i_{t-1}} \cap A_{k+1}), \qquad (16.5) \end{aligned}$$

where the sum in (16.5) indexed by $k+1$ is

$$\sum_{1<2<\cdots<k} n(A_1 \cap \cdots \cap A_k \cap A_{k+1}) = n(A_1 \cap \cdots \cap A_k \cap A_{k+1}).$$

Let us examine the terms in (16.5). For $t = 1$ we let

$$C_1' = \sum_{1 \le i \le k} n(A_i) \quad \text{and} \quad C_1 = \sum_{1 \le i \le k+1} n(A_i).$$

By (16.3), C_1' is the sum of the integers $n(A_i)$ in $n(B_k)$, so that

$$C_1 = C_1' + n(A_{k+1}). \qquad (16.6)$$

For a given integer $2 \le t \le k$, we let C_t' be the sum of all integers $n(A_{i_1} \cap \cdots \cap A_{i_t})$ such that $A_{i_1}, \cdots, A_{i_t}$ is a list of t sets taken from the set $\{A_1, \ldots, A_k\}$ and for which $1 \le i_1 < \cdots < i_t \le k$. Specifically,

$$C_t' = \sum_{1 \le i_1 < \cdots < i_t \le k} n(A_{i_1} \cap \cdots \cap A_{i_t}),$$

and by substituting C_t' into (16.3) we see that

$$n(B_k) = \sum_{t=1}^{k} (-1)^{t-1} C_t'. \qquad (16.7)$$

Similarly, we let

$$\begin{aligned} C_t &= \sum_{1 \le i_1 < \cdots < i_t \le k+1} n(A_{i_1} \cap \cdots \cap A_{i_t}) \\ &= C'_t + \sum_{1 \le i_1 < \cdots < i_{t-1} \le k} n(A_{i_1} \cap \cdots \cap A_{i_{t-1}} \cap A_{k+1}). \end{aligned} \tag{16.8}$$

Next, given an integer $1 < t \le k$, let C''_t denote the sum over all of the integers $n(A_{i_1} \cap \cdots \cap A_{i_{t-1}} \cap A_{k+1})$ that are determined by $t-1$ sets $A_{i_1}, \cdots, A_{i_{t-1}}$ taken from the set $\{A_1, \ldots, A_k\}$ and for which $1 \le i_1 < \cdots < i_{t-1} \le k$. Thus, for $1 < t \le k$ we have

$$\begin{aligned} C''_t &= \sum_{1 \le i_1 < \cdots < i_{t-1} \le k} n(A_{i_1} \cap \cdots \cap A_{i_{t-1}} \cap A_{k+1}), \text{ and} \\ C''_{k+1} &= n(A_1 \cap \cdots \cap A_k \cap A_{k+1}). \end{aligned}$$

Subsequently,

$$n(B_k \cap A_{k+1}) = \sum_{t=1}^{k+1} (-1)^{t-1} C''_t. \tag{16.9}$$

Given $1 < t \le k$, the definition (16.8) of C_t shows us that

$$\begin{aligned} C_t &= \sum_{\{1 \le i_1 < \cdots < i_t \le k+1\}} n(A_{i_1} \cap \cdots \cap A_{i_t}) \\ &= \sum_{\{1 \le i_1 < \cdots < i_t \le k\}} n(A_{i_1} \cap \cdots \cap A_{i_t}) \\ &\qquad - (-1) \sum_{\{1 \le i_1 < \cdots < i_{t-1} \le k\}} n(A_{i_1} \cap \cdots \cap A_{i_{t-1}} \cap A_{k+1}) \\ &= C'_t - (-1) C''_t. \end{aligned}$$

Thus,

$$C_t = C'_t - (-1) C''_t \tag{16.10}$$

for each integer $1 < t \le k$. Furthermore, given $t = k+1$, then it is natural to let $C'_{k+1} = 0$, so that

$$C_{k+1} = n(A_1 \cap \cdots \cap A_k \cap A_{k+1}) = C'_{k+1} - (-1) C''_{k+1}.$$

Finally, let $N_{k+1} = n(B_{k+1})$. Then by (16.4),

$$N_{k+1} = n(B_k) + n(A_{k+1}) - n(B_k \cap A_{k+1}).$$

Because $C'_{k+1} = 0$, and by the equations (16.7) and (16.9), we see that

$$\begin{aligned} N_{k+1} &= \sum_{t=1}^{k+1}(-1)^{t-1}C'_t + n(A_{k+1}) - (-1)^{t-2}\sum_{t=2}^{k+1}C''_t \\ &= (C'_1 + n(A_{k+1})) + (-1)^{t-1}\sum_{t=2}^{k+1}(C'_t - (-1)C''_t) \end{aligned}$$

which from (16.6) and (16.10) satisfies

$$N_{k+1} = C_1 + (-1)^{t-1}\sum_{t=2}^{k+1}C_t = (-1)^{t-1}\sum_{t=1}^{k+1}C_t.$$

This last sum for $N_{k+1} = n(B_{k+1})$ and the sum (16.7) for $n(B_k)$ are the same except for the use of $k+1$. Then by mathematical induction, the boxed formula in Theorem 16.4.1 is True for *every integer* $n \geq 2$. This completes the proof.

EXAMPLE 16.4.2 We know the inclusion/exclusion formula for $n = 3$ sets from Chapter 3, so let's proceed by assuming that $n = 4$. Then by Theorem 16.4.1,

$$\begin{aligned} n(A_1 \cup A_2 \cup A_3 \cup A_4) &= \sum_{t=1}^{4}(-1)^{t-1}C_t \\ &= n(A_1) + n(A_2) + n(A_3) + n(A_4) \\ &\quad -n(A_1 \cap A_2) - n(A_1 \cap A_3) - n(A_1 \cap A_4) \\ &\quad\quad -n(A_2 \cap A_3) - n(A_2 \cap A_4) \\ &\quad +n(A_1 \cap A_2 \cap A_3) + n(A_1 \cap A_2 \cap A_4) \\ &\quad\quad +n(A_1 \cap A_3 \cap A_4) + n(A_2 \cap A_3 \cap A_4) \\ &\quad -n(A_1 \cap A_2 \cap A_3 \cap A_4). \end{aligned}$$

16.5 A Common Miscount

A formula for derangements that is commonly found on the internet gives us a chance to examine another erroneous form of counting, thus proving again that an idea may not be right even though it is used by a large number of people. Let us build this formula up from ground level in the manner in which it is given in most internet files and postings. We will then point out the error that is employed in the counting.

THEOREM 16.5.1 *Let $n \geq 2$ be an integer, and let F_n be the number of permutations of $\{1, \ldots, n\}$ that fix at least one element of $\{1, \ldots, n\}$. There is an integer $N \in \mathbb{N}$ such that*

$$0 < F_n - n!(1 - e^{-1}) < 1 \text{ for each integer } n > N.$$

Proof: A given permutation π of $\{1, \ldots, n\}$ is said to fix i if $\pi(i) = i$. Given $i \in \{1, \ldots, n\}$, let P_i be *the set of permutations* π that fix at least i. Then, given an integer $1 \leq t \leq n$ and a sequence $1 \leq i_1 < \cdots < i_t \leq n$ of integers, the set

$$P_{i_1} \cap \cdots \cap P_{i_t}$$

consists of *the permutations that fix at least* $i_1, \ldots, i_t$. Then

$$F_n = \sum_{t=1}^{n} \left((-1)^{t-1} \sum_{1 \leq i_1 < \cdots < i_t \leq n} n(P_{i_1} \cap \cdots \cap P_{i_t}) \right). \tag{16.11}$$

Fix an integer $1 \leq t \leq n$, choose a subset $\{i_1, \ldots, i_t\}$ of $\{1, \ldots, n\}$ in exactly $\begin{pmatrix} n \\ t \end{pmatrix}$ ways, which is the number of expressions $P_{i_1} \cap \cdots \cap P_{i_t}$.

Since any permutation of $\{1, \ldots, n\}$ that fixes $\{i_1, \ldots, i_t\}$ induces a permutation of the $n - t$ elements in the complement of $\{i_1, \cdots, i_t\}$ in $\{1, \ldots, n\}$, $n(P_{i_1} \cap \cdots \cap P_{i_t}) = (n - t)!$.

After arranging the i_j as the sequence $1 \le i_1 < \cdots < i_t \le n$, we find that

$$\sum_{1\le i_1<\cdots<i_t\le n} n(P_{i_1} \cap \cdots \cap P_{i_t}) = (n-t)! \cdot \binom{n}{t}$$

After substituting into (16.11) we see that

$$\begin{aligned} F_n &= \sum_{t=1}^{n} \left((-1)^{t-1} \sum_{1\le i_1<\cdots<i_t\le n} n(P_{i_1} \cap \cdots \cap P_{i_t}) \right) \\ &= \sum_{t=1}^{n} (-1)^{t-1} (n-t)! \cdot \binom{n}{t} \\ &= (-1)n! \sum_{t=1}^{n} (-1)^t \frac{1}{t!} \\ &= (-1)n! \left(\sum_{t=0}^{n} (-1)^t \frac{1}{t!} - 1 \right) \end{aligned}$$

Since $e^x = \sum_{t=0}^{\infty} \frac{x^t}{t!}$ and since $t! = t \cdots (n+1)n!$ for each $t > n$ we can further write

$$\begin{aligned} F_n &= (-1)n! \left(e^{-1} - 1 - \sum_{t=n+1}^{\infty} (-1)^t \frac{1}{t!} \right) \\ &= (-1)n! \left(e^{-1} - 1 - \sum_{t=n+1}^{\infty} (-1)^t \frac{1}{t \cdots (n+1)n!} \right). \quad (16.12) \end{aligned}$$

Because the infinite series $\sum_{t=0}^{\infty} \frac{(-1)^t}{t!}$ is absolutely convergent, we can actor $\frac{1}{n!}$ from (16.12) to form

$$F_n = (-1)n! \left(e^{-1} - 1 - \frac{1}{n!} \sum_{t=n+1}^{\infty} (-1)^t \frac{1}{t \cdots (n+1)} \right)$$

$$
\begin{aligned}
&= (-1)\frac{n!}{n!}\left(n!e^{-1} - n! - \sum_{t=n+1}^{\infty}(-1)^t\frac{1}{t\cdots(n+1)}\right) \\
&= (-1)\left(n!(e^{-1} - 1) - \sum_{t=n+1}^{\infty}(-1)^t\frac{1}{t\cdots(n+1)}\right) \\
&= \sum_{t=n+1}^{\infty}(-1)^t\frac{1}{t\cdots(n+1)} + n!(1 - e^{-1}). \qquad (16.13)
\end{aligned}
$$

The Alternating Series Test shows that $\sum_{t=n+1}^{\infty}(-1)^t\frac{1}{t\cdots(n+1)}$ converges, so there is a least integer $N > 0$ such that

$$0 < \sum_{t=m}^{\infty}(-1)^t\frac{1}{t\cdots m} < 1$$

for every integer $N \leq m$. Then *for large enough* $n \geq N$,

$$F_n = \sum_{t=n+1}^{\infty}(-1)^t\frac{1}{t\cdots(n+1)} + n!(1 - e^{-1})$$

$$0 < F_n - n!(1 - e^{-1}) = \sum_{t=n+1}^{\infty}(-1)^t\frac{1}{t\cdots(n+1)} < 1$$

by our choice of N. Hence

$$0 < F_n - n!(1 - e^{-1}) < 1. \qquad (16.14)$$

This completes the proof.

REMARK 16.5.2 Now here is the error found in most calculations of F_n of this type. This inequality holds only for integers $n > N$. The inequality (16.14) does not hold for integers $n < N$ because

$$\sum_{t=n+1}^{\infty}(-1)^t\frac{1}{t\cdots(n+1)} \geq 1$$

for all integers $n < N$.

COROLLARY 16.5.3 *Let $N > 0$ be the least integer such that $\sum_{t=N} (-1)^t \frac{1}{t!} < 1$, let $n > N$ be an integer, and let D_n be the number of derangements of $\{1, \ldots, n\}$. Then*

$$0 < D_n - n!e^{-1} < 1$$

Proof: Choose $n > N$. Since F_n is the number of permutations fixing at least one element of $\{1, \ldots, n\}$, we have $D_n = n! - F_n$. By Theorem 16.5.1, there is an inequality (16.14), so

$$\begin{aligned} 0 < D_n &= n! - F_n < 1 + n!e^{-1} \\ 0 < D_n - n!e^{-1} &= n! - F_n - n!e^{-1} < 1. \end{aligned}$$

This completes the proof.

16.6 Exercises

1. Find F_6. Answer: $6(5! - 86) + \binom{6}{2}(4! - 15) + \binom{6}{3}(3! - 4) + \binom{6}{4}(2! - 1) + 1$

2. Find F_7. Answer: Apply the formula with $n = 7$.

3. Find the number of derangements of $\{1, 2, 3, 4, 5, 6\}$. Answer: $6! - F_6$.

4. Find the number of derangements of $\{1, 2, 3, 4, 5, 7\}$.

5. Find the number of ways that 5 men can leave the house party with someone else's coat. Answer: $5! - F_5$. Locate F_5 in this chapter.

6. Find the number of ways that 6 men can leave the house party with someone else's coat.

Chapter 17

Probability Vocabulary

There are a number of mathematics books written for college freshmen claiming to discuss the probability of equally likely outcomes in a couple of chapters. Their treatment of the subject is based mostly on an intuitive understanding about the basics of the subject. Most of the discussions in these chapters lack a firm mathematical grounding. Some of these chapters even show that their discussions come to the authors purely from an intuitive understanding of the subject. While intuition is necessary to solve problems in a subject, it should not be used as the initial explanation as to why certain formulas are true.

In this chapter we will present what we think is a modern approach to probability and its problems. The emphasis is on random processes and equally likely outcomes, which will be defined later. As we did in the rest of the book, we will try to justify any probability that we present. Only after a mathematical definition is given will we use intuition to understand a concept. The problems will use our knowledge of counting problems and a knowledge of defined terms. These mathematical definitions seem to agree with our intuitive understanding of what probability entails.

17.1 Vocabulary

An *experiment* is something you do, and an *outcome* is what we observe at the end of an experiment. The set of all outcomes of an experiment is called the *sample space.* We usually use S to denote the sample space of an experiment. An *event* is a subset of S, the sample space. Thus, if E and F are events in S, we can talk about $E \cup F$, (the event that E or F occurs), $E \cap F$ (the event that E and F occur), and E' (the event that E does not occur).

What we need now are some examples. We will use these examples many times as the chapter continues.

EXAMPLE 17.1.1 Experiment: Flip a coin and observe the side facing up. The side of the coin is either denoted as H or T. It does not make any difference which side you mark as H, it only matters that you make the identification consistently. The same side should be called H at all times. The outcomes for this experiment are H and T, so the sample space is $\mathrm{S} = \{H, T\}$. The coin is *fair.* The outcomes H and T occur with the same frequency in many numbers of flips. That is, if you flip the coins many times (more than millions of times) and observe the outcome of each flip, then the fraction

$$\frac{\text{number of } H\text{'s observed}}{\text{number of } T\text{'s observed}}$$

will be very close to 1. More flips produce a better approximation to 1 in this fraction. Thus, with many flips of the coin, an H should be observed as many times as a T is observed. Since no one can flip a coin this often, this experiment must take place in that Platonic universe of mathematical ideas and shapes.

EXAMPLE 17.1.2 A die is a cube (made of plastic or ivory if you need a physical model), with numbers 1 through 6 stamped into its sides. Experiment: Toss a die and observe the top face when the die stops. The sample space of the experiment is $\mathrm{S} = \{1, 2, 3, 4, 5, 6\}$. This is just the set of numbers on the faces of the die. The die is *fair.* The outcomes occur with the same frequency in many numbers of tosses. We mean by this that if $k \in \mathrm{S}$ then over many tosses of the die,

$$\frac{\text{number of tosses that } k \text{ is observed}}{\text{number of tosses of the die}}$$

is approximately $\frac{1}{6}$. More tosses of the die will make a better approximation of $\frac{1}{6}$. Thus, a 3 is observed about once in every 6 tosses. But do not expect 3 to turn up every 6 tosses of the die. This approximation is only good for large numbers of tosses.

EXAMPLE 17.1.3 Let B be a bag of distinguishable objects. Experiment: Choose an object from B and observe the chosen object. Then $\mathrm{S} = \{x \mid x \text{ is an object in } B\}$. Notice you cannot write down the elements exactly, but they do make a set. We assume that the objects in B are chosen with the same frequency in many numbers of choices.

We can put these simple experiments to work for us, and use them to generate more interesting sample spaces.

EXAMPLE 17.1.4 Let $n > 0$ be a whole number. Suppose you have an experiment with sample space S, and you execute the experiment n times in sequence. This represents a new experiment that has sample space

$$\mathrm{S}_n = \{x \mid x \text{ is a word consisting of exactly } n \text{ symbols from the alphabet } \mathrm{S}\}.$$

Since each place in the n letter word can be filled in $n(\mathrm{S})$ ways, $n(\mathrm{S}_n) = n(\mathrm{S})^n$.

For instance, let $\mathrm{S} = \{H, T\}$. If $n = 1$ then $\mathrm{S}_1 = \mathrm{S}$ and $n(\mathrm{S}_1) = 2^1$.

If $n = 2$ then

$$\mathrm{S}_2 = \{xy \mid x, y \in \mathrm{S}\} = \{HH, HT, TH, TT\}.$$

Notice that $n(\mathrm{S}_2) = 2^2$.

If $n = 10$ then

$$\mathrm{S}_{10} = \{\mathbf{x} \mid \mathbf{x} \text{ is a 10-letter word from the alphabet } \{H, T\}\}.$$

Since each place in the 10-letter word can be filled in 2 ways, $n(\mathrm{S}_{10}) = 2^{10}$.

Sometimes, it is not convenient to write S_n as words with n letters. It could be, for instance, that the elements of the alphabet

S do not look like letters or symbols at all. In this case, a word with n letters from S is replaced by a sequence of n symbols from S.

$$\mathrm{S}_n = \{(x_1, \ldots, x_n) \mid x_1, \ldots, x_n \in \mathrm{S}\}.$$

Then $n(\mathrm{S}_n) = n(\mathrm{S})^n$. We will try to use the sample space that best fits our problem and our intuitive understanding of the symbols in S.

EXAMPLE 17.1.5 Let $n > 0$ be a whole number. Let $\mathrm{S} = \{H, T\}$. Experiment: Toss a coin n times and observe the outcome of the sequence of H's and T's. The sample space is

$$\mathrm{S}_n = \{\mathbf{x} \mid \mathbf{x} \text{ is an } n \text{ letter word from the alphabet } \{H, T\}\}.$$

From Example 17.1.4, $n(\mathrm{S}) = 2^n$.

EXAMPLE 17.1.6 Let $\mathrm{S} = \{1, 2, 3, 4, 5, 6\}$ = the sample space of throwing a die once. Experiment: Toss a die twice. The sample space is $\mathrm{S}_2 = \{(x, y) \mid x, y \in \{1, 2, 3, 4, 5, 6\}\}$. Since there are 6 ways to fill x and y individually, the multiplication principal shows us that $n(\mathrm{S}_2) = 6^2 = 36$.

EXAMPLE 17.1.7 Let us look again at the dice experiment. Experiment: Throw 2 identical fair dice and observe the top faces of the dice.

There seems to be some confusion as to the proper sample space for this experiment. It is preferable to make the outcomes of the experiment equally likely. That is, over many executions of the experiment, the outcomes occur with the same frequency.

The sample space $\{2, 3, 4, 5, 6, 7, 8, 9, 10, 11, 12\}$ with 11 outcomes has been suggested, but this is entirely incorrect. The experiment observes the dots on the top faces, and not their sum. Furthermore, the 11 outcomes are not equally likely. For instance, 2 does not occur very often, while 7 occurs quite often. Thus, this is not the sample space for the dice experiment.

We will argue that a much larger sample space is in order. To begin with, assume that one of the dice is colored white and the other is colored black. You and a friend are playing a game with the dice. Because you see the black and white dice, you suggest that the sample space should be the following array of pairs. The first entry of a pair is the upper face observed on the black die. The second entry of a pair is the upper face observed on the white die.

$$\text{Pairs} = \left\{ \begin{array}{cccccc} (1,1) & (1,2) & (1,3) & (1,4) & (1,5) & (1,6) \\ (2,1) & (2,2) & (2,3) & (2,4) & (2,5) & (2,6) \\ (3,1) & (3,2) & (3,3) & (3,4) & (3,5) & (3,6) \\ (4,1) & (4,2) & (4,3) & (4,4) & (4,5) & (4,6) \\ (5,1) & (5,2) & (5,3) & (5,4) & (5,5) & (5,6) \\ (6,1) & (6,2) & (6,3) & (6,4) & (6,5) & (6,6) \end{array} \right\}$$

There is some sense to this since you can tell the difference between the black die and the white die. Furthermore, since the dice are fair, these pairs represent equally likely outcomes in the experiment.

Now comes the twist. Your friend is blind. He does not see the difference in the dice. He only knows what you are telling him. The sample space of this experiment for your friend should be the same as the sample space produced by your sighted observations of the outcomes of this experiment. Thus, we can assume that if the dice are indistinguishable then the sample space for the experiment of throwing 2 fair dice will be the array of pairs *Pairs.* Thus, $n(\text{Pairs}) = 36$.

REMARK 17.1.8 But here is the most compelling reason that the sample space should be the set *Pairs* above. When you hold those "identical" dice in your hand, *you see two dice.* You observe two different dice in your hand, not one. If these dice were identical then they would be indistinguishable. You would therefore see only one die in your hand. You see two dice because *they are not identical.* Thus, the sample space for the dice experiment should be the above set called *Pairs*, which consists of 36 different pairs whose entries are between 1 and 6.

EXAMPLE 17.1.9 Let S be the sample space of tossing one fair die and observing its upper face. Then S_2, the sample space of tossing the one die twice, and *Pairs*, the sample space of tossing a pair of dice once, are the same set. Therefore the game of tossing one die twice, and two dice once, have the same probabilities associated with them. This is two games with the same sample spaces and the same probabilities of the outcomes.

EXAMPLE 17.1.10 Suppose you have a bag full of n different chips. Experiment: Pull a handful of chips from the bag and note the chips chosen. It is usually difficult to make a list of all possible outcomes, but we can do this.

$$\text{S} = \{\mathbf{x} \mid \mathbf{x} \text{ is a subset of the chips in the bag}\}$$

It is important that we count S. The number of subsets of a set of n elements is 2^n. Thus, $n(\text{S}) = 2^n$.

Chapter 18

Equally Likely Outcomes

The probability that we will examine here is the probability of experiments with *equally likely outcomes*. (Definition to follow below.) That describes all of the examples that we gave in Section 17.1, so we can examine the probability of those experiments. In almost all of our examples it will be clear what the experiment is, so we will stop writing Experiment: unless it adds significantly to the discussion. You cannot do probability without a sample space so in each problem we must write down S.

Firstly, let us define probability. A *probability function* is a function Pr on the outcomes in a sample space S with the following two properties:

1. $0 \leq \Pr(x) \leq 1$ for all $x \in \mathrm{S}$, and

2. If we add up all of the values $\Pr(x)$ over all $x \in \mathrm{S}$ then the sum is 1.

We will verify that the probability function we choose satisfies this definition. A sample space S consists of *equally likely outcomes* if there is a number $0 \leq p \leq 1$ such that $\Pr(x) = p$ for all $x \in \mathrm{S}$.

18.1 Outcomes in Experiments

EXAMPLE 18.1.1 The sample space for flipping a fair coin is $\mathrm{S} = \{H, T\}$. As we stated in Example 17.1.1, over many flips of the coin, the number of observed H's is approximately the number of observed T's. This appears to be a pair of equally likely outcomes, and there are 2 of them so we *define* a *probability function*, or just a *probability*, by writing

$$\Pr(H) = \Pr(T) = \frac{1}{2}.$$

We might also have written

$$\Pr(\mathbf{x}) = \frac{1}{2} \text{ for each } \mathbf{x} \in \mathrm{S}.$$

EXAMPLE 18.1.2 Toss a fair die once. The sample space is $\mathrm{S} = \{1, 2, 3, 4, 5, 6\}$. As we found in Example 17.1.2, a given $k \in \mathrm{S}$ occurs about $\frac{1}{6}$ of the time. This is what it means for S to consist of equally likely outcomes. Thus, we *define* a probability on S as follows.

$$\Pr(k) = \frac{1}{6} \text{ for all } k \in \mathrm{S}.$$

EXAMPLE 18.1.3 This is the general case. Repeat an experiment E with sample space S n times. This number n will have to be a large number. If we observe that there is a fraction $0 \leq p \leq 1$ such that

$$\frac{\text{number of observed occurrences of } \mathbf{x}}{n(\mathrm{S}_n)} \text{ is approximately } p$$

for all $\mathbf{x} \in \mathrm{S}$, then we say that S consists of *equally likely outcomes.* In this case, we *define* the probability on S by writing

$$\Pr(\mathbf{x}) = p \text{ for all } \mathbf{x} \in \mathrm{S}.$$

That is, every outcome in S has probability p *by definition.*

We keep returning to the word *definition* or *define*, so let me explain. There is no magic probability function out there that must be drawn to your sample space. If you choose a finite sample space

S with *equally likely outcomes* then the most natural of probability functions on S is

$$\Pr(\mathbf{x}) = \frac{1}{n(\mathrm{S})} \text{ for each } \mathbf{x} \in \mathrm{S}.$$

But it is not the only one. This formula applies only to those S in which the outcomes $\mathbf{x}$ are equally likely. If the $\mathbf{x}$ are not equally likely, then $\frac{1}{n(\mathrm{S}))}$ does not generally describe the probability of $\mathbf{x}$. For example, if we toss two dice and use the sample space of pairs then

$$\Pr(x, y) = \frac{1}{36} \text{ for each } (x, y) \in \mathrm{S}$$

because the outcomes in S are equally likely. That is what we mean by fair dice. However, if we choose the sample space $B = \{2, 3, 4, 5, 6, 7, 8, 9, 10, 11, 12\}$ for the experiment of tossing a pair of dice then

$$\Pr(2) = \frac{1}{36}$$

because the only way to toss a 2 is to toss the pair $(1, 1)$. On the other hand,

$$\Pr(3) = \frac{2}{36} = \frac{1}{18}$$

since 3 can be rolled as either pair $(2, 1)$ or $(1, 2)$. If you are using dice of the same color to conduct this experiment then try using the black and white dice used in Example 17.1.7. Obviously, the outcomes in B are not equally likely.

Now, let us see whether this satisfies our definition of probability function. Suppose that S is a sample space consisting of equally likely outcomes, and let $n(\mathrm{S})$ be finite. Then for any $\mathbf{x} \in XX$,

$$0 \leq \Pr(\mathbf{x}) = \frac{1}{n(\mathrm{S})} \leq 1$$

since $\Pr(x)$ is a fraction of positive numbers, and since $n(S) \geq 1$. Add up all of the values $\Pr(\mathbf{x})$. Since each value $\Pr(\mathbf{x})$ is just $\frac{1}{n(\mathrm{S})}$, the sum over all $\mathbf{x} \in \mathrm{S}$ is

$$n(\mathrm{S}) \cdot \frac{1}{n(\mathrm{S})} = 1.$$

Thus, $\Pr(\mathbf{x})$ satisfies our conditions and so we call $\Pr(\mathbf{x})$ a probability function.

We will need to know the probability of an event so suppose the sample space S consists of equally likely outcomes. Let E be an event in S. Then we *define*

$$\Pr(E) = \frac{n(E)}{n(\mathrm{S})}.$$

EXAMPLE 18.1.4 1. Toss a fair coin and observe the outcome. The fairness of the coin shows that that H and T are equally likely outcomes. Since the sample space consists of exactly 2 outcomes, the probability that you toss an H is $\Pr(H) = \frac{1}{2}$.

2. Toss a fair die and observe the value of the top face. The fairness of the die shows us that the outcomes are equally likely. Well, we could hardly call the die fair if one of the outcomes occurs more often than another. The sample space has exactly 6 outcomes, so $\Pr(5) = \frac{1}{6}$. The probability that we toss an odd numbered face is $\Pr(\text{odd number}) = \frac{n(\{1,3,5\})}{6} = \frac{1}{2}$.

3. Toss 2 dice and observe the two top faces. The sample space S of equally likely outcomes contains exactly 36 outcomes, so $\Pr(2,1)$ $=$ the probability of throwing $(2,1) = \frac{1}{36}$. Find the probability that we roll a 7. We count that there are 6 pairs in S that add up to 7. Hence, $\Pr(7) = \frac{6}{36} = \frac{1}{6}$.

EXAMPLE 18.1.5 Toss a coin 10 times and observe the sequence of the 10 occurrences of H's and T's. Find the probability that none of the flips yields an H.

Solution: The sample space S is the set of all 10-letter words from the alphabet of $\{H, T\}$. We have done this problem often. $n(\mathrm{S}) = 2^{10}$. The only outcome in S without an H is the word $TTTTTTTTTT$. Thus, $\Pr(TTTTTTTTTT) = \dfrac{1}{2^{10}}$.

EXAMPLE 18.1.6 Toss a coin 10 times and observe the sequence of the 10 occurrences of H's and T's. Find the probability that the word $\mathbf{x}$ has exactly 2 H's.

Solution: The sample space S has $n(\mathrm{S}) = 2^{10}$ equally likely outcomes. The word $\mathbf{x}$ is formed by choosing the 2 places of 10 places for the H, and then filling in the remaining places with T. The event $E = \{\text{words } \mathbf{x} \,|\, \mathbf{x} \text{ contains exactly 2 } H\text{'s}\}$ satisfies $n(E) = \binom{10}{2}$. Thus,

$$\Pr(x) = \frac{n(E)}{n(\mathrm{S})} = \frac{1}{2^{10}}\binom{10}{2}.$$

The *complement formula* for probability is derived as follows. Assume the outcomes in S are equally likely, and let E be an event in S. Consider the event E. Since $E \cap E' = \emptyset$, the inclusion/exclusion principle shows us that

$$\begin{aligned} n(\mathrm{S}) &= n(E \cup E') \\ &= n(E) + n(E') - n(E \cap E') \\ &= n(E) + n(E') - n(\emptyset) \\ &= n(E) + n(E'). \end{aligned}$$

Hence,

$$1 = \Pr(E \cup E') = \frac{n(E \cup E')}{n(\mathrm{S})} = \frac{n(E) + n(E')}{n(\mathrm{S})}$$

$$= \frac{n(E)}{n(\mathrm{S})} + \frac{n(E')}{n(\mathrm{S})} = \Pr(E) + \Pr(E')$$

and therefore

$$\boxed{\Pr(E) = 1 - \Pr(E').}$$

EXAMPLE 18.1.7 Toss a coin 10 times and observe the sequence of the 10 occurrences of H's and T's. Find the probability that the word $\mathbf{x} \in \mathrm{S}$ has at least 1 H.

Solution: We use the complement formula. The sample space S satisfies $n(\mathrm{S}) = 2^{10}$.

Let E be the set of words $\mathbf{x} \in \mathrm{S}$ such that $\mathbf{x}$ contains at least 1 H. Then E' is the set of words $\mathbf{x}$ that do not contain an H. Observe that $E' = \{TTTTTTTTTT\}$. We calculated $\Pr(E') = \frac{1}{2^{10}}$ in Example 18.1.5, so $\Pr(E) = 1 - \Pr(E') = 1 - \frac{1}{2^{10}}$.

The above example shows that our substantial experience with counting problems will be used in solving some interesting probability problems.

We have encountered the *standard deck of cards* when we counted poker hands.

EXAMPLE 18.1.8 Choose 1 card from a standard deck of cards. What is the probability that the card is an A, or an A?

Solution: The sample space S is the set of cards in the deck. There are 52 cards in S. Let E be the event that we choose an A. That is, E is set of A. Then $n(E) = 4 =$ the number of As in S, and so $\Pr(E) = \frac{n(E)}{n(\mathrm{S})} = \frac{4}{52}$.

Here is the punch line:

$$\Pr(E) = \frac{4}{52} = \frac{8}{104} = .07692308$$

to 8 decimal places. But when you tune into that cable channel and watch the experts play the game, the probability the program gives you when a player has to draw an A from the deck is 8% or .08. (This assumes that no one else has an A.) There are two possible explanations for this difference. One is that they ignored the 4 in 104. I think that is unlikely. More likely is that they rounded their probability to two decimal places and turned the actual value .07692308 into .08. Certainly 8% is more understandable to their audience than a 7 decimal number, so this explanation is credible.

However, when you see the amount of money that these players are betting, you wonder why someone doesn't complain. Over a long period of games, you would bet assuming that the probability that you draw an A is 8%, which it is not. The reality is that you would draw an A only about 7.692308% of the time. The difference between 8% and the actual probability, about .3%, measures the number of times that you bet with bad information. If they were given this incorrect information, someone would be denied a substantial amount of money over many games. That's a hanging offense in Texas!

EXAMPLE 18.1.9 Draw five cards from the standard deck. Find the probability that you draw 2 K's and 3 Q's, or equivalently, 2 K's and 3 Q's.

Solutions: The sample space S is all 5 card subsets of the standard deck. This is a combination problem, so $n(\mathrm{S}) = \binom{52}{5}$.

Construct an outcome from the event E by first choosing 2 of 4 K's and then choose 3 of 4 Q's. This is done in $n(E) = \binom{4}{2}\binom{4}{3}$ ways. Thus,

$$\Pr(E) = \frac{n(E)}{n(\mathrm{S})} = \frac{\binom{4}{2}\binom{4}{3}}{\binom{52}{5}}.$$

EXAMPLE 18.1.10 Draw 5 cards from a standard deck. Find the probability that you draw 2 of one kind, 2 of another kind, and a third kind of card. This is called drawing 2 pair.

Solution: We begin with $n(\mathrm{S}) = \binom{52}{5}$.

Let E be the event that we draw 2 pair. Construct an outcome $\mathbf{x} \in E$. You will draw two kinds to be used in $\mathbf{x}$, and there is no way to give them an order in the hand. Choose 2 of 13 kinds in $\binom{13}{2}$ ways. Then choose 2 of 4 cards from each kind in $\binom{4}{2}^2$. Finally, choose the fifth card from the 44 cards whose kinds have not been chosen. This is done in 44 ways. The total number of hands in E is $n(E) = 44 \cdot \binom{13}{2}\binom{4}{2}^2$. Thus,

$$\Pr(E) = \frac{44 \cdot \binom{13}{2}\binom{4}{2}^2}{\binom{52}{5}}.$$

18.2 Exercises

You possess a bag of 40 chips. Ten are red round chips, 10 are red square chips, 10 are black round chips, and 10 are black square chips. Each of these type of chips is numbered 1 through 10. E is the event given in the problems below. To help with the other problems count the number of ways can you choose a handful of 10 chips and find $\Pr(E)$. Answer: $n(\mathcal{U}) = \binom{40}{10}$ and $\Pr(E) = n(E)/n(\mathcal{U})$.

1. Find the probability that you chose 10 red chips. Answer: $E =$ the event you choose 10 red chips. $n(E) = \binom{20}{10}$

2. Find the probability that you chose 6 red round chips and 4 black round chips. Answer: $E =$ you choose 6 red round chips and 4 black round chips. $n(E) = \binom{10}{6}\binom{10}{4}$.

3. Find the probability that you choose exactly 6 red round chips, and 4 other chips. Answer: $E =$ you choose exactly 6 red round chips, and 4 other chips. $n(E) = \binom{10}{6}\binom{30}{4}$.

4. Find the probability that you choose at least 1 red chip. Answer: $E =$ you choose at least 1 red chip. $n(E) = n(\mathcal{U}) - \binom{20}{10}$.

5. Find the probability that you choose at least 2 red round chips. Answer: $E =$ you choose at least 2 red round chips. $n(E) = n(\mathcal{U}) - \binom{30}{10} - 10 \cdot \binom{30}{9}$.

6. Find the probability that you choose at least 1 red round chip and at least 1 black round chip. Answer: $E =$ you choose at least 1 red round chip and at least 1 black round chip. $n(E) = n(\mathcal{U}) - 2 \cdot \binom{30}{10} + \binom{20}{10}$.

7. Find the probability that you choose 3 chips numbered by 5, 3 chips numbered by 6, 3 chips numbered by 7, and a chip not numbered by 5, 6, or 7. Answer: $E =$ you choose 3 chips numbered by 5, 3 chips numbered by 6, 3 chips numbered by 7, and a chip not numbered by 5, 6, or 7. $n(E) = 4^3 \cdot 28$.

8. Find the probability that you choose a sequence of 8 chips. Answer: $n(E) = 2 \cdot 4^{10}$.

Chapter 19

Probability Trees

Suppose you draw a card from a deck and remove it. The sample space for this experiment is the set of cards in the standard deck. Then you draw a second card. The probability you drew a K, or a King, on the first draw is $4/52$. The probability you drew a Q, or a Queen, on the second draw given that you drew a King on the first try is $4/51$. The probability of the second draw is *dependent* upon the first draw. The multiplication principle still holds so the probability of drawing a K, removing a K, and then drawing a Q is

$$\frac{4}{52}\frac{4}{51}.$$

This is the kind of probabilistic problem that we consider in this chapter.

19.1 Tree Diagrams

A better name for these problems would be *tree diagram problems* since the best visual device for doing these problems is a decision tree. We will label the edges (lines) in the tree with probability values, and we will label the vertices (dots) in the tree with outcomes. The probabilities are what make these problems interesting. In general, the probability on these branches will be different as they are each calculated in different sample spaces. For this reason, you must be cautious when labeling these trees.

Also, the tree can be used to calculate the probability of a series of events. By multiplying or adding the probabilities from different edges you find the probability for sometimes interesting and applicable events. Unfortunately, you must fill in every numerical aspect of the tree before you can calculate probabilities with certainty. Some examples will explain this, but first we need a little more language.

Two events E and F are *mutually exclusive* if $E \cap F = \emptyset$. Given mutually exclusive events E and F, the inclusion/exclusion principle shows us that

$$\Pr(E \cap F) = \Pr(\emptyset) = 0.$$

Hence,

> If E and F are *mutually exclusive events* then
> $\Pr(E \cup F) = \Pr(E) + \Pr(F)$.

We can extend this formula to three mutually exclusive events as follows. Suppose that E, F, and G are *pairwise mutually exclusive events.* That is, suppose that E and F are mutually exclusive events, that E and G are mutually exclusive events, and that F and G are mutually exclusive events. All pairs of events are mutually exclusive. Then

$$(E \cup F) \cap G = (E \cap G) \cup (F \cap G) = \emptyset \cup \emptyset = \emptyset$$

since E and G are mutually exclusive events, and since F and G are mutually exclusive events. Thus, $E \cup F$ and G are mutually exclusive events. Hence,

$$\Pr(E \cup F \cup G) = \Pr((E \cup F) \cup G) = \Pr(E \cup F) + \Pr(G).$$

Since E and F are mutually exclusive

$$\Pr(E \cup F) + \Pr(G) = \Pr(E) + \Pr(F) + \Pr(G).$$

Therefore,

> If E, F, and G are *pairwise mutually exclusive events*
> $\Pr(E \cup F \cup G) = \Pr(E) + \Pr(F) + \Pr(G)$

EXAMPLE 19.1.1 A bag contains 5 red chips, 6 white chips, and 7 blue chips. Choose a chip and observe its color, do not replace it in the bag, and then choose another chip and observe its color. What is the probability that you chose 1 red and then 1 white?

Solution: We set up a tree to illustrate the experiment's outcomes. Here is a completely detailed tree for this experiment. Each vertex represents a choice of chips. The 3 arrows from each vertex represent the 3 colors of the chips. The edges represent the probability that we choose that color. We need to do some elementary calculations before we can label the tree with probabilities.

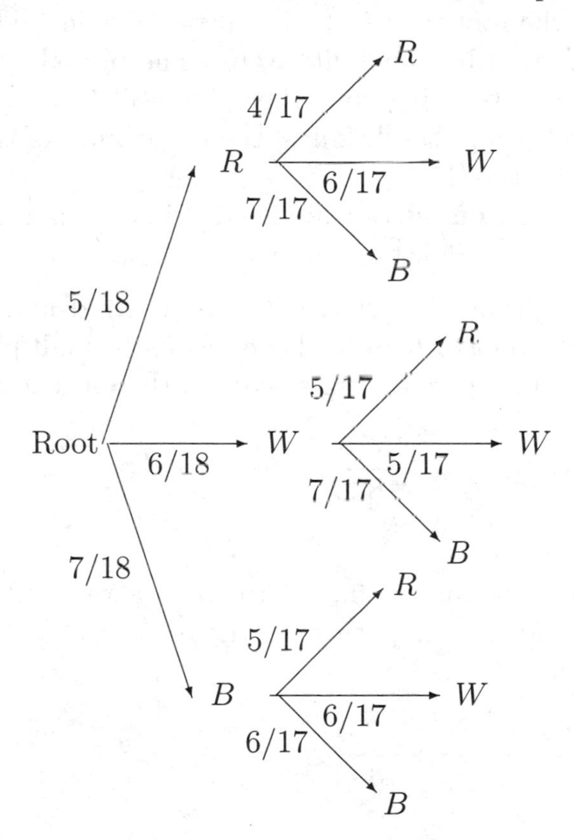

For the first choice there are 18 chips to choose from so the sample space has 18 chips in it. The event R that you choose a red chip has probability $\Pr(R) = 5/18$. The event W that you choose a white chip has probability $\Pr(W) = 6/18$. The event B that you choose a blue chip has probability $\Pr(B) = 7/18$.

For the second choice there are 17 chips (because you did not replace the first choice). The probability $\Pr(R)$ depends on what the first choice was. The tree keeps track of this kind of detail quite nicely.

Let me do one such edge to begin the explanation. Suppose you are looking at edges at the top of the diagram. The first edge shows that you choose a red chip in R, and then chose a red chip again. The first probability is $5/18$. You remove one chip from R making a set of 4 red chips in a larger set of 17 chips. So the probability in the second edge is $4/17$.

The path from the root to R to W is a little different. The edge from the root to R is still probability $5/18$. The probability from R to W is found by observing that there are still 6 white chips, and there are only 17 chips total. Thus, the probability of the edge connecting R and W is $6/17$.

The rest of the edges on the tree are filled in in a similar manner. Now to use the tree.

1. What is the probability that you choose a red and then a red? Follow the path root to R to R. Then, using the multiplication principle the probability of choosing a red and then a red is

$$\frac{5}{18}\frac{5}{17}.$$

2. What is the probability that you choose a red and then a white chip? Again, follow the path from the root to R to W. The probability we want is then

$$\frac{5}{18}\frac{6}{17}.$$

3. To find the probability that you chose a red chip as your second chip, find all paths that terminate on R. By the multiplication principle , the probabilities along the paths are

$$\frac{5}{18}\frac{4}{17}, \quad \frac{6}{18}\frac{5}{17}, \text{ and } \frac{7}{18}\frac{5}{17}.$$

This is the probability of a red and a red being chosen (RR), or a white and a red being chosen (WR), or a blue and a red being chosen (BR). We then have to calculate $\Pr(RR \cup WR \cup BR)$. But the events RR and WR are mutually exclusive since you cannot choose a red chip and a white chip on the first draw. Similarly, RR and BR are mutually exclusive, and WR and BR are mutually exclusive. The inclusion/exclusion formula then shows us that

$$\Pr(RR \cup WR \cup BR) = \Pr(RR) + \Pr(WR) + \Pr(BR).$$

Hence,

$$\Pr(\text{a red chip is your second chip}) = \frac{5}{18}\frac{4}{17} + \frac{6}{18}\frac{5}{17} + \frac{7}{18}\frac{5}{17}.$$

4. The probability that you chose at least 1 red chip is found by finding all of the paths that contain R and then add the products they represent as follows:

$$\Pr(\text{at least 1 red chip}) = \frac{5}{18} + \frac{6}{18}\frac{5}{17} + \frac{7}{18}\frac{5}{17}.$$

The reason for adding the probabilities corresponding to each path is that two paths in the tree represent mutually exclusive events. In this example, $\frac{5}{18}$ comes from the path containing R and R. You chose a red chip on your first try. Call this event R. Furthermore, $\frac{6}{18}\frac{5}{17}$ comes from the path that contains W and R. You choose a white and then a red chip. Call this event WR. The events R and WR that these paths represent are mutually exclusive because in one you chose a red first, and in the other you chose a white first. Similarly, the path corresponding to a choice of a blue chip and then a red chip is mutually exclusive with the other paths R and WR. Call this event BR.

Thus, by the inclusion/exclusion principle,

$$\begin{aligned}\Pr(\text{at least 1 red chip}) &= \Pr(R \cup WR \cup BR)\\ &= \Pr(R) + \Pr(WR) + \Pr(BR)\\ &= \frac{5}{18} + \frac{6}{18}\frac{5}{17}. + + \frac{7}{18}\frac{5}{17}.\end{aligned}$$

The above example featured a problem with 3 different decisions to be made at each juncture. Suppose we only have 2 decisions at each juncture. Then the tree we draw has only 2 edges at each juncture. The next example shows us how such a tree is special.

EXAMPLE 19.1.2 There is a bag of 10 black chips and 15 white chips. You choose 1 chip, remove it from the bag, and then choose another chip. What is the probability that you will choose at least 1 black chip?

Solution: The tree we will construct looks like this.

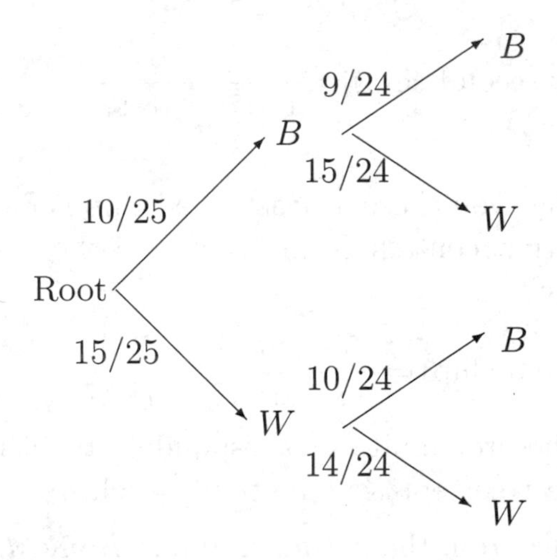

Label the diagram using B to indicate a black chip and W to indicate a white chip. The sample space of choices of one chip has 25 chips in it. The probability that you choose a black chip on the first choice is $\frac{10}{25}$ and the probability that you choose a white chip is $\frac{15}{25}$. The probability that you choose a black chip on the second draw, given that you already drew a black chip on the first draw is $\frac{9}{24}$. The probability that you choose a white chip on the second draw, given that you drew a black chip on the first draw is $\frac{9}{24}$. The other edges of the graph are determined in the same manner.

We give some related problems.

1. The probability that you drew a black and then a black chip is found by following the path from Root to B and then to B. Call

this event B. By the multiplication principle , the probability is

$$\Pr(BB) = \frac{10}{25}\frac{9}{24}.$$

2. The probability that you drew a black chip on your second draw is determined by finding all paths that lead to a B. Call this event B. Use the inclusion/exclusion formula to conclude that

$$\Pr(B) = \frac{10}{25}\frac{9}{24} + \frac{15}{25}\frac{10}{24}.$$

3. The probability that you drew at least 1 black chip is found by adding up the probabilities associated with each path that contains a B.

$$\Pr(\text{at least 1 black}) = \frac{10}{25} + \frac{15}{25}\frac{10}{24}.$$

EXAMPLE 19.1.3 For the purposes of this problem, 60% of the population is infected with bald cat disease. A test is available to decide with 90% assurance that you have or do not have the disease. Let I denote an *infected* person and NI denote a person who is *not infected.* P denotes a positive test for the disease, and N denotes a negative test result. The associated tree looks like this.

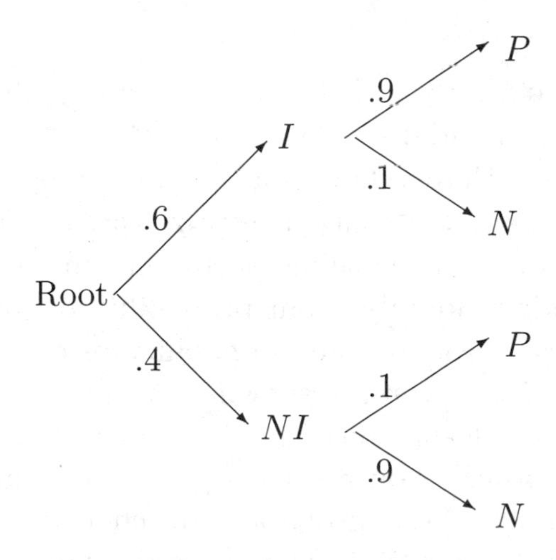

Let me explain the switch of the values on the edges containing NI and P. If you are infected, I, then your test result will be positive 90% of the time because the test is accurate 90% of the time. If you are not infected, NI, then your test result will be negative 90% of the time since the test is accurate 90% of the time.

Let us answer some questions about the test. You are chosen at random from the total population.

1. The probability that you get a positive result is found by finding paths that end in P. As in the previous examples, different paths produce mutually exclusive events. Then by the inclusion/exclusion principle ,

$$\Pr(P) = (.6)(.9) + (.4)(.1) = .58.$$

2. Find the probability that you get a positive result if you are not infected. To find this probability you find the path corresponding to a noninfected person, NI, and then find the path corresponding to the positive test result. The resulting probability is is $(.4)(.1) = .04$. Thus, 4 out of 100 people will be told that they are infected when they are not infected. This is called a *false positive test.*

The following is an interesting problem in mathematics as well as in American culture.

EXAMPLE 19.1.4 The Monty Hall Problem. Monty Hall was a showman who in the 1960's and 1970's was Master of Ceremonies for a television game show. Here is how you play his game.

You are given 3 curtains. One contains a prize (a winning curtain), and the other two contain nothing (losing curtains). Your choice of a curtain is made randomly. Your probability of choosing a winning curtain (W) is 1/3, and your probability of choosing a losing curtain (L) is 2/3. Next, you are shown that one of the curtains you did not choose is a losing curtain. This leaves 2 unknown curtains, yours and one other. You are then given the chance to choose from these 2 curtains. Your probability of choosing a winning curtain (W) on this choice is 1/2. The probability of choosing a losing curtain at this time is 1/2.

There are some popular questions associated with this problem that can be answered if you keep a rational head on your shoulders. Otherwise, you may loose sight of the object of this game.

The tree diagram associated with the Monty Hall Problem is below.

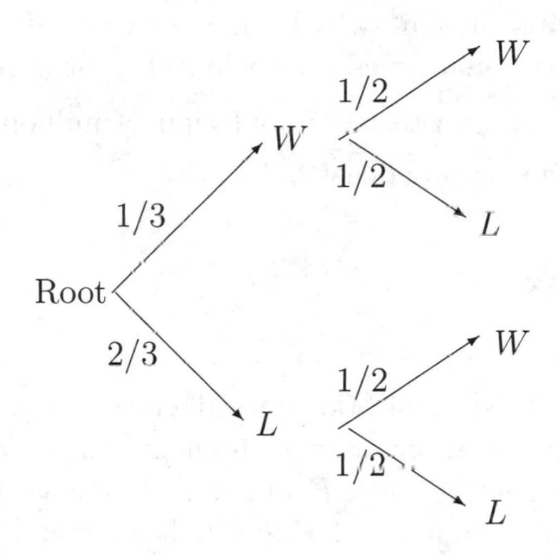

1. What is the probability that you win? To win, your choices end in a W. Take the paths that end in a W and use the multiplication principle and the inclusion/exclusion formula to find that

$$\Pr(\text{winning}) = \frac{1}{3}\frac{1}{2} + \frac{2}{3}\frac{1}{2} = \frac{1}{2}.$$

This is why Monty thought he was giving the people a fair game.

2. What is the probability that you win without switching your curtain in the second choice? To win without switching, you had to choose the winner W to begin with and then stand pat. By following the path from Root to W to W we get probability

$$\Pr(\text{win without switching}) = \frac{1}{3}\frac{1}{2} = \frac{1}{6}.$$

3. What is the probability that you chose W after switching from your first choice. To win after switching you had to have a

loser L in your first choice, and then switch into the winner. The probability for this is

$$\Pr(\text{win after switching}) = \frac{2}{3}\frac{1}{2} = \frac{1}{3}.$$

The answers to questions 2 and 3 show that you are twice as likely to win in this game if you switch when you are given the chance. However, these probabilities do not effect your game because the probability $\frac{1}{2}$ applies to the population of millions that play the game each time it is broadcast.

19.2 Exercises

Suppose that 80% of the student body passed a test administered by some university. Due to an error in technology, the computer grading the exam assigned the pass P and fail F grades to the exams *randomly*. So half were marked P and half were marked F. Choose an exam, note how the student did on the exam, then note how the computer graded the exam.

1. Draw the associated tree.
2. Trace out the path on the tree corresponding to the event that the computer gives you a passing grade given that you passed the test. What is that probability? Answer: $.8 \cdot .5$
3. Trace out the path on the tree corresponding to the event that the computer gives the student a deserving grade. (That is, a grade to match your actual test score.) What is that probability? Answer: .5
4. Trace out the path on the tree that corresponds to the event that the computer gave you an undeserving grade. What is that probability? Answer: .5

You possess a bag of 40 chips. Ten are red round chips, 10 are red square chips, 10 are black round chips, and 10 are black square chips. Each type of chip is numbered 1 through 10. Choose a chip, note its color and shape (round or square), remove it from the bag, choose another chip, and note its shape.

5. Draw the associated tree.

6. What is the probability that you choose a round chip?
Answer: $\frac{20}{40}$

7. What is the probability that you choose a round chip given that you chose a black chip? Answer: $\frac{10}{40}\frac{10}{39} + \frac{10}{40}$

8. What is the probability that you choose at least 1 round chip?
Answer: $1 - 2 \cdot \left(\frac{10}{40}\frac{10}{39} + \frac{10}{40}\right)$

Chapter 20

Independent Events

Let S be a sample space and let E and F be events in S. Recall that E and F are *mutually exclusive* if $E \cap F = \emptyset$. And if E and F are mutually exclusive events, then the inclusion/exclusion principle shows us that

$$\Pr(E \cup F) = \Pr(E) + \Pr(F).$$

Dually, we say that E and F are *independent events* if

$$\boxed{\Pr(E \cap F) = \Pr(E)\Pr(F).}$$

Notice the difference. If E and F are *mutually exclusive* events then Pr changes $E \cup F$ into addition. If E and F are *independent* events then Pr changes $E \cap F$ into multiplication.

20.1 Independence

EXAMPLE 20.1.1 Given an experiment with sample space S of equally likely outcomes, and given any event $E \subset \mathrm{S}$ then E and S are independent events. To see this, calculate $\Pr(E \cap \mathrm{S})$ as follows. Notice that $E \cap \mathrm{S} = E$ and that $\Pr(\mathrm{S}) = 1$. Then

$$\Pr(E \cap \mathrm{S}) = \Pr(E) = \Pr(E) \cdot 1 = \Pr(E)\Pr(\mathrm{S}).$$

This is what we had to show to prove that E and S are independent.

EXAMPLE 20.1.2 Let S be a finite sample space of experiment Exp1. Let T be a finite sample space of experiment Exp2. It could happen that Exp1 = Exp2.

A new experiment is defined by doing experiment Exp1 and then doing experiment Exp2. The new experiment is denoted Exp1 $\times$ Exp2. The sample space of Exp1$\times$Exp2 is $\mathrm{S} \times \mathrm{T}$, which has $n(\mathrm{S} \times \mathrm{T}) = n(\mathrm{S})n(\mathrm{T})$ outcomes.

Let $E \subset \mathrm{S}$ be an event. The event that E occurs in the execution of Exp1 in the experiment Exp1$\times$Exp2 is the set of pairs

$$E \times \mathrm{T} = \{(e,t) \mid e \in E \text{ and } t \in \mathrm{T}\}.$$

You might say colloquially that $E \times \mathrm{T}$ is the event that we produce an outcome of E in the execution of Exp1 in Exp1$\times$Exp2. For example, if $E = \{H\}$ and if Exp1 is to flip a coin, then $E \times \mathrm{T}$ is the event that you tossed an H in Exp1. You don't care what the second part Exp2 of the experiment is.

Let $F \subset \mathrm{T}$ be an event. The event that F occurs in the execution of Exp2 in the experiment Exp1$\times$Exp2 is the set of pairs

$$\mathrm{S} \times F = \{(s,f) \mid s \in \mathrm{S} \text{ and } f \in F\}.$$

You might say colloquially that $\mathrm{S} \times F$ is the event that we produce an outcome of F in the execution of Exp2 in Exp1$\times$Exp2. For example, if $F = \{Q\}$ and if Exp2 is to draw a card, then $\mathrm{S} \times F$ is the event that you drew a Q in Exp2. You don't care what the first part Exp1 of the experiment is.

EXAMPLE 20.1.3 Let S be a finite sample space of an experiment Exp1. Let T be a finite sample space of an experiment Exp2. Let $E \subset \mathrm{S}$ and $F \subset \mathrm{T}$ be events. Then $E \times \mathrm{T}$ and $\mathrm{S} \times F$ are events in $\mathrm{S} \times \mathrm{T}$. We will show that $E \times \mathrm{T}$ and $\mathrm{S} \times F$ are independent events.

To prove this rather interesting claim, we need to show that the two events satisfy our equation.

We have $n(E \times \mathrm{T}) = n(E)n(\mathrm{T})$ and $n(\mathrm{S} \times F) = n(F)n(\mathrm{S})$. Then

$$\Pr(E \times \mathrm{T}) = \frac{n(E)n(\mathrm{T})}{n(\mathrm{S})n(\mathrm{T})} = \frac{n(E)}{n(\mathrm{S})}$$

and

$$\Pr(\mathrm{S} \times F) = \frac{n(F)n(\mathrm{S})}{n(\mathrm{S})n(\mathrm{T})} = \frac{n(F)}{n(\mathrm{T})}.$$

Thus,

$$\Pr(E \times \mathrm{T})\Pr(\mathrm{S} \times F) = \frac{n(E)}{n(\mathrm{S})}\frac{n(F)}{n(\mathrm{T})} = \frac{n(E)n(F)}{n(\mathrm{S})n(\mathrm{T})}.$$

Next, consider a pair $(x, y) \in (E \times \mathrm{T}) \cap (\mathrm{S} \times F)$. Because $(x, y) \in E \times \mathrm{T}$, $x \in E$, and because $(x, y) \in \mathrm{S} \times F$, $y \in F$. Thus,

$$(E \times \mathrm{T}) \cap (\mathrm{S} \times F) = \{(x, y) \mid x \in E \text{ and } y \in F\} = E \times F,$$

so that

$$n((E \times \mathrm{T}) \cap (\mathrm{S} \times F)) = n(E \times F) = n(E)n(F).$$

Hence,

$$\Pr((E \times \mathrm{T}) \cap (\mathrm{S} \times F)) = \frac{n(E)n(F)}{n(\mathrm{S} \times \mathrm{T})} = \frac{n(E)n(F)}{n(\mathrm{S})n(\mathrm{T})}.$$

But then

$$\Pr((E \times \mathrm{T}) \cap (\mathrm{S} \times F)) = \frac{n(E)n(F)}{n(\mathrm{S})n(\mathrm{T})} = \Pr(E \times \mathrm{T})\Pr(\mathrm{S} \times F)$$

by the boxed formula above. Because they satisfy the equation of the definition, we have proved that $E \times \mathrm{T}$ and $\mathrm{S} \times F$ are independent events.

EXAMPLE 20.1.4 Let S be the sample space of the experiment Exp. Form a new experiment Exp×Exp. This is the experiment of doing Exp once, replacing the outcome, and then doing Exp again. The sample space is $\mathrm{S} \times \mathrm{S}$. Let E and F be any events in S. Then by Example 20.1.3, the events $E \times \mathrm{S}$ and $\mathrm{S} \times F$ are independent events in $\mathrm{S} \times \mathrm{S}$.

Notice the lack of hypotheses on E and F. This paucity of assumptions gives us a wealth of independent events.

EXAMPLE 20.1.5 Let $\mathrm{S} = \{H, T\}$ = the sample space for the experiment of flipping a coin. The outcomes in S are equally likely, and $n(\mathrm{S}) = 2$. Toss a fair coin twice and observe the outcomes. The sample space is $\mathrm{S} \times \mathrm{S} = \{(x, y) \mid x, y \in \mathrm{S}\}$ and $n(\mathrm{S} \times \mathrm{S}) = 4$.

Let $E = \{H\}$ and let $F = \{T\}$. Then the event that we first flip an H is $E \times \mathrm{S}$, and the event that we flip a T on the second flip is $\mathrm{S} \times F$. Observe that $n(E \times \mathrm{S}) = n(E)n(\mathrm{S}) = 2 = n(\mathrm{S} \times F)$.

Then by Example 20.1.3, $E \times \mathrm{S}$ and $\mathrm{S} \times F$ are independent events in $\mathrm{S} \times \mathrm{S}$. Specifically,

$$\Pr(E \times \mathrm{S}) \Pr(\mathrm{S} \times F) = \frac{2}{4}\frac{2}{4} = \frac{1}{4} = \Pr(E \cap F).$$

This shows that our first toss of an H is not influenced *mathematically* by the second toss.

Intuitively, this says that flipping an H on the first flip, and flipping a T on the second flip are independent events. Our intuition is inaccurate in this case. We must specify the sample space before we can determine that we have independent events, or that we have events at all.

20.2 Logical Consequences of Influence

Some have argued that since the second toss does *not influence* the first, the events E are independent. This is backwards thinking, though, and in error. The error is that we do not know what the term *influence* means mathematically, and until we know that fact, the use of *influence* to conclude mathematical properties of an event is in error. In mathematics, terms are defined precisely, or they are not used. The correct argument goes like this. First you show that E and F satisfy the equation of independence, and then you *intuitively* conclude that the first toss is not influenced by a second toss.

EXAMPLE 20.2.1 There is some confusion in the literature as to what our definition of *independent events* entails and why *influence* is not precise enough to prove independence. These two examples will illustrate my point.

1. Toss a fair coin and observe the outcomes. The sample space is $S = \{H, T\}$ and it consists of equally likely outcomes. Let E be the event that you flip an H, and let F be the event that you flip a T. Then $E \cap F = \{H\} \cap \{T\} = \emptyset$ so that $\Pr(E \cap F) = 0$. Since $\Pr(E) = \Pr(F) = \frac{1}{2}$,

$$\Pr(E \cap F) = 0 \neq \frac{1}{2}\frac{1}{2} = \Pr(E)\Pr(F).$$

According to our definition of independent events, E and F *are not independent events.* That does not mean, however, that E and F influence each other physically. Indeed, we do not know what it means for E *to influence* F because as yet, no one has posited a good definition of what it means for E *to influence* F.

When we say that two events are independent we mean that they satisfy a certain equation. They are not independent events until we show that they satisfy that equation. Throwing an H and throwing a T are not independent events in $S = \{H, T\}$ because they fail to satisfy our equation $\Pr(E \cap F) = \Pr(E)\Pr(F)$. Before we examine them, the definition also requires that the events E and F come from the same sample space.

2. Draw a card from a deck and replace it. The sample space S is the set of all cards in the deck, and so $n(S) = 52$. Let E be the event that we drew a K and let F be the event that we drew a Q. Let us examine the independence of drawing a K and drawing a Q.

Observe that $n(E) = n(F) = 4$ since there are exactly 4 K's in the deck and exactly 4 Q's in the deck. Obviously, $E \cap F = \emptyset$ because a K is not a Q. Hence,

$$\Pr(E)\Pr(F) = \frac{4}{52}\frac{4}{52} \neq 0 = \Pr(E \cap F)$$

so that E and F *are not independent events.* They do not satisfy our equation so they do not qualify as independent events. Does that mean that drawing a K *influences* your draw of a Q? We do not know, since no one seems to know the definition of *influence.*

Here is an example about drawing cards, influence, and independence.

EXAMPLE 20.2.2 Draw a card from a standard deck, and replace it. The sample space S is the set of cards, and $n(\mathrm{S}) = 52$. Suppose you draw 2 cards in a row with replacement. The sample space is $\mathrm{S} \times \mathrm{S} = \{(x, y) \mid x, y \text{ is a card}\}$. The outcomes in $\mathrm{S} \times \mathrm{S}$ are equally likely, and $n(\mathrm{S} \times \mathrm{S}) = 52^2$.

The event that we first choose a K is $E = \{(K, x) \mid x \in \mathrm{S}\}$, and the event that we choose a Q in the second draw is $F = \{(x, Q) \mid x \in \mathrm{S}\}$. Then $n(E) = n(\mathrm{S}) = 52 = n(F)$. By Example 20.1.3, E and F are independent events in $\mathrm{S} \times \mathrm{S}$. Our use of a Q here to define F is not special. We would have had the same outcome for the equation if we used any other card. We conclude that the draw of a K in the first draw is independent of the second draw.

Some have used backwards thinking on cards. They assume that the second card does *not influence* the first card drawn, and then they would conclude independence. The error in the argument is that we do not know precisely what *influence* means. The correct argument is that the events E and F are independent events, and so *intuitively*, the K in the first draw is not influenced by the second draw. But no one can use an intuitive term like *influence* to conclude *independence.*

Here is the most damning of reasons for ignoring *influence* in all of probability.

EXAMPLE 20.2.3 Suppose that we have events $E = \{H\}$ which is the flip of a coin, and $F = \{6\}$ which is the toss of a die. Ignore the fact that we have violated the hypotheses that we have laid down for independence by taking events E an F from unequal sample spaces. Independent events must come from the same sample space and the same experiment.

The pseudo-intellectual argument states that coins and dice do not influence each other so E and F must be *independent.* While none of us would call E and F *dependent* events, neither are they independent. When we calculate probabilities we find that

$$\Pr(E)\Pr(F) = \frac{1}{12} \neq 0$$

while

$$\Pr(E \cap F) = 0$$

because obviously $E \cap F = \emptyset$. Failure to satisfy the formula shows us that E and F are *not independent* events.

Therefore, this use of *influence* produces three logical possibilities. Events can be independent or dependent, or as is the case for coins and dice, *not independent.* This is impossible, so we must abandon the use of *influence* in calculating the independence of events.

EXAMPLE 20.2.4 Recall the sample space S for tossing a pair of dice is a set of 36 pairs. See Example 17.1.7.

Are the events that you throw and $(1, 4)$ and a $(2, 5)$ independent events? The sample space is the set S of pairs a faces on the dice, the outcomes in the sample space are equally likely, and $n(\mathrm{S}) = 36$. Let $E = \{(1, 4)\}$ and $F = \{(2, 5)\}$. Then, $E \cap F = \emptyset$ so that

$$\Pr(E)\Pr(F) = \frac{1}{36}\frac{1}{36} \neq 0 = \Pr(E \cap F).$$

Hence, despite what you may practice at a casino, the tosses $(1, 4)$ and $(2, 5)$ are *not independent events.*

EXAMPLE 20.2.5 In the game of dice there are certain tosses that will allow you to continue tossing the dice, and there are certain tosses that keep you from throwing the dice again. What these pairs are is unimportant, unless you want to make money from this mathematics, which is a dirty thought. So onto the problem. You toss a pair of dice once, and then again if possible. The sample space for this experiment is the set of pairs of pairs

$$\mathrm{S} = \{((x, y), (u, v)) \mid x, y, u, v \text{ are faces on the dice}\},$$

and so $n(\mathrm{S}) = 36^2$.

Let E be the event that you first toss the specific pair $(1,4)$. This allows you another toss. Let F be the event that you toss the specific pair $(2,5)$ on the second toss. Then E is the set of pairs of pairs $\{((1,4),(u,v))|u,v \text{ are faces on the dice}\}$, and then F is the set of pairs of pairs $\{((x,y),(2,5))|x,y \text{ are faces on the dice}\}$. Hence, $E \cap F = \{(1,4),(2,5)\}$. We conclude that $n(E) = n(F) = 36$, and $n(E \cap F) = 1$. It follows that

$$\Pr(E)\Pr(F) = \frac{36}{36^2}\frac{36}{36^2} = \frac{1}{36^2} = \Pr(E \cap F).$$

Therefore, the event of throwing a $(1,4)$ on your first throw is independent of your toss of a $(2,5)$ on the second throw.

Indeed, there is nothing special about $(2,5)$ in this problem. The problem could have been any specific pair (U,V) and the results would have been the same. Thus, your toss of a $(1,4)$ on the first throw is independent of your second toss *no matter what that second throw is.*

The danger of using a word in your mathematics without first knowing its precise definition is that you can unknowingly change the meaning of the word without knowing it. A rather extreme case of this abuse is illustrated nicely in the following story.

EXAMPLE 20.2.6 Let X be the number of legs on a horse. A horse has 2 hind legs, and up front he has his fore legs. These two and four legs make 6 legs. Then $X = 6$ is an even number. But 6 is an odd number of legs for a horse to have, so X is odd. Thus, X is even and odd. The only number that is both even and odd is infinity, so X is infinity. Thus, a horse has an infinite number of legs.

This little bit of arithmetical nonsense is what happens when we do not precisely define and use our terms in mathematics.

20.3 Exercises

1. There are independent events E and F such that $\Pr(E) = .3$ and $\Pr(F) = .2$. Fill in the Venn diagram whose regions are labeled with probabilities of E, F, $E \cap F$ and the other regions.

Refer to exercises 2 through 6. A bag holds 13 black numbered chips and 13 white numbered chips. The black chips are numbered 1 to 13 and the white chips are numbered 1 to 13. In an experiment you choose and remove a chip from the bag. You note its number and color. Then you choose another chip, noting its number and color.

2. What is the sample space? Answer: $S = \{(x, y) \mid x \neq y$ are chips from the bag$\}$. Notice that $n(S) = 26 \cdot 25$.

3. Let $x \neq y$ be specific different numbered and colored chips. Let E be the event you choose x on the first draw, and let F be the event you choose y on a second draw. Write down the events E and F as pairs in S. Answer: $E = \{(x, b) \mid b$ is a chip from the bag and $x \neq b\}$.

4. Find $n(E)$, $n(F)$, $\Pr(E)$, and $\Pr(F)$. Answer: $n(E) = 25 = n(F)$, $\Pr(E) = \dfrac{1}{26} = \Pr(F)$.

5. Find $E \cap F$ and $\Pr(E \cap F)$. Answer: $E \cap F = \{(x, y)\}$, $\Pr(E \cap F) = \dfrac{1}{26 \cdot 25}$.

6. Show that E and F are not independent.

Refer to exercises 7 through 10. A bag holds 13 black numbered chips and 13 white numbered chips. The black chips are numbered 1 to 13 and the white chips are numbered 1 to 13. In an experiment you choose and discard a chip from the bag. You note its number and color. Then you choose and discard another chip from the bag. You note its number and color. You choose yet another chip from the bag, noting its number and color.

7. What is the sample space and how many outcomes are in it? Answer: $S = \{(a, b, c) \mid a, b, c$ are chips in the bag, $a \neq b$ and $a, b \neq c\}$. $n(S) = 26 \cdot 25 \cdot 24$.

8. Let $x \neq y$ be specific different numbered and colored chips. Write down the events E that you choose x on the first draw, and F that you choose y on a third draw. Answer: $E = \{(x, b, c) \mid b$ and c are chips in the bag such that $b \neq x$ and $c \neq x, b\}$

9. Find $n(E)$, $n(F)$, $\Pr(E)$, $\Pr(F)$. Answer: $n(E) = 25 \cdot 24 = n(F)$ $\Pr(E) = \dfrac{1}{26} = \Pr(F)$.

10. Find $E \cap F$ and show that E and F are not independent. Answer: $E \cap F = \{(x, b, y) \mid b \text{ is a chip in the bag, } x \neq b \text{ and } y \neq b, x\}$, $n(E \cap F) = 24$, $\Pr(E \cap F) = \dfrac{1}{26 \cdot 25}$.

Refer to exercises 11 and 12. In an experiment you choose and discard a card from a standard deck, and then you choose another card.

11. What is the sample space?

12. Draw any card A and then any card B in this experiment. Show that these events are not independent.

Chapter 21

Sequences and Probability

The multiplication principle has been used throughout the first 19 chapters of this book. Our use here will be a little more systematic. For example, suppose that you have events $E_1, \ldots, E_n$ for some whole number $n > 0$, and that we wish to know the probability of the sequential event $E =$

$$E_1 \text{ and then } E_2 \ldots \text{ and then } E_n.$$

Also suppose that each event E_i occurs with the same probability, namely $\Pr(E_i) = p$ for each whole number $i = 1, \ldots, n$. Then, by the multiplication principle, the probability that the sequential event E occurs is

$$p^n.$$

We will investigate variations to this analysis in this chapter.

21.1 Sequences of Events

A sequence of events can sometimes be analyzed for its probability, and there results a probability with a simple formula. Here are a few examples.

A list of events $E_1, \ldots, E_n$ is said to be a list of *pairwise independent events* if any two events E_i and E_j on this list are independent events. In other words, if E and F are events on this list then E and F are independent events. *Given a list* $E_1, \ldots, E_n$ *of pairwise independent events*, an iterated use of the definition of independent events shows us that

$$\Pr(E_1 \cap \cdots \cap E_n) = \Pr(E_1) \cdots \Pr(E_n).$$

EXAMPLE 21.1.1 People pass a series of 10 windows, and they open or close them randomly. After some time, it is determined that the probability that a window is open at the end of the day is $p = \frac{1}{10}$. Furthermore, it is determined that the open windows are pairwise independent events.

1. Find the probability that each of the windows is open at the end of the day.

Solution: For $i = 1, \ldots, 10$, let E_i denote the event that window number i is open. Then, $E_1 \cap \cdots \cap E_{10}$ is the event that window 1 is open, and window 2 is open, ..., and window E_{10} is open. In other words, $E_1 \cap \cdots \cap E_{10}$ is the event that *each window is open.* Then our assumption is that $E_1, \ldots, E_{10}$ is a list of pairwise independent events. Thus,

$$\begin{aligned} \Pr(\text{each window is open}) &= \Pr(E_1 \cap \cdots \cap E_{10}) \\ &= \Pr(E_1) \cdots \Pr(E_{10}) \\ &= \frac{1}{10} \cdots \frac{1}{10} = \frac{1}{10^{10}}. \end{aligned}$$

2. Find the probability that at least 1 window is closed.

Solution: The complement of the condition *at least 1 window is closed* is the condition that *which each window is open.* Then, by the complement formula and part 1 we have

$$\begin{aligned} \Pr(\text{at least 1 window is closed}) &= 1 - \Pr(\text{each window is open}) \\ &= 1 - \frac{1}{10^{10}}. \end{aligned}$$

EXAMPLE 21.1.2 Consider 30 switches. It is determined that the probability that a switch is off at noon is $p = .4$. Number the switches from 1 to 30. For each $i = 1, \ldots, 30$, let F_i be the event that switch i is on at noon. Assume that $F_1, \ldots, F_{30}$ is a list of pairwise independent events.

1. Find the probability that at least one switch is off at noon.

Solution: The complement of the condition *at least one switch is off* is *no switch is off*, or equivalently, *each switch is on.* By assumption, the probability that switch i is on at noon is $\Pr(F_i) = .6$. Then, because $F_1, \ldots, F_{30}$ is a list of pairwise independent events, the probability that each switch is on at noon is

$$\begin{aligned}\Pr(F_1 \cap \cdots \cap F_{30}) &= \Pr(F_1) \cdots \Pr(F_{10}) \\ &= \underbrace{.6 \cdots .6}_{30} = .6^{30}.\end{aligned}$$

Then,

$$\begin{aligned}\Pr(\text{at least one switch is off}) &= 1 - \Pr(\text{each switch is on}) \\ &= 1 - .6^{30}.\end{aligned}$$

2. We observe that F_i' is the event that *switch i is off at noon.* Assume that the events $F_1', \ldots, F_{30}'$ form a list of pairwise independent events. Find the probability that some switch is on at noon.

Solution: The complement of *some switch is on at noon* is *no switch is on at noon*, or equivalently, *each switch is off at noon.* Then the event

$$F_1' \cap \cdots \cap F_{30}' = \text{the event that each switch is off at noon.}$$

The probability that switch i is on at noon is $\Pr(F_i) = .6$, so the probability that switch i is off at noon is $\Pr(F_i') = 1 - .6 = .4$. Hence,

$$\begin{aligned}\Pr(\text{each switch is off at noon}) &= \Pr(F_1' \cap \cdots \cap F_{30}') \\ &= \Pr(F_1') \cdots \Pr(F_{30}') \\ &= .4^{30}.\end{aligned}$$

The complement formula then shows us that

$$\begin{aligned}\Pr(\text{some switch is on at noon}) &= 1 - \Pr(\text{each switch is off at noon}) \\ &= 1 - .4^{30}.\end{aligned}$$

Now let's get a little more abstract. Here is how we can analyze the experiment of repeating an experiment several times.

EXAMPLE 21.1.3 Let Exp be an experiment with sample space SS, and fix the important event E in SS. We will let Exp_n denote the experiment of repeating Exp sequentially n times. The sample space for Exp_n is

$$SS_n = \underbrace{SS \times \cdots \times SS}_{n}.$$

Choose an element $(\mathbf{x}_1, \ldots, \mathbf{x}_n)$ is SS_n and consider $\mathbf{x}_i$. Let E_i be the event that $\mathbf{x}_i \in E$. Then,

$$E_i = SS \times \cdots \times \underbrace{E}_{i\text{-th place}} \times \cdots \times SS.$$

By the multiplication principle

$$\begin{aligned}
\Pr(E_i) &= \Pr(SS) \cdots \Pr(E) \cdots \Pr(SS) \\
&= 1 \cdots \Pr(E) \cdots 1 \\
&= \Pr(E).
\end{aligned}$$

You might say intuitively that E_i is the event that E occurs is the i-th place in $(\mathbf{x}_1, \ldots, \mathbf{x}_n)$. One can argue as we did in Example 20.1.4 that $E_1, \ldots, E_n$ is *a list of pairwise independent events.*

Furthermore, let E_i' be the event

$$E_i' = SS \times \cdots \times \underbrace{E'}_{i\text{-th place}} \times \cdots \times SS.$$

Then $E_1', \ldots, E_n'$ is *a list of pairwise independent events.*

EXAMPLE 21.1.4 Take the setting from the previous example. Choose a sequence $(\mathbf{x}_1, \ldots, \mathbf{x}_n) \in SS_n$. Find the probability that some $\mathbf{x}_i$ is in E'.

Solution: Let $\Pr(E) = p$ be the probability that E occurs. Observe that the event *some* $\mathbf{x}_i$ *is in* E' is the same as the event *some*

$\mathbf{x}_i$ *is not in* E. Its complement is *each* $\mathbf{x}_i$ *is in* E, and this predicate defines E_i. Then

$$\begin{aligned}\{\mathbf{x} \in SS \,|\, \text{all } \mathbf{x}_i \in E\} &= \{\mathbf{x} \,|\, \mathbf{x}_1 \in E\} \text{ and } \cdots \text{ and } \{\mathbf{x} \,|\, \mathbf{x}_n \in E\} \\ &= E_1 \cap \cdots \cap E_n.\end{aligned}$$

By Example 21.1.3, $E_1, \ldots, E_n$ is a list of pairwise independent events, so

$$\begin{aligned}\Pr(\text{each } \mathbf{x}_i \text{ is in } E) &= \Pr(E_1 \cap \cdots \cap E_n) \\ &= \underbrace{\Pr(E) \cdots \Pr(E)}_{n} \\ &= p^n.\end{aligned}$$

Therefore, by the complement formula,

$$\begin{aligned}\Pr(\text{some } \mathbf{x}_i \text{ is in } E') &= 1 - \Pr(\text{each } \mathbf{x}_i \text{ is in } E) \\ &= 1 - p^n\end{aligned}$$

EXAMPLE 21.1.5 Take the setting from Example 21.1.3. Choose a sequence $(\mathbf{x}_1, \ldots, \mathbf{x}_n) \in SS_n$. Find the probability that some $\mathbf{x}_i$ is in E.

Solution: As before, let $\Pr(E) = p$ be the probability that E occurs. Then $\Pr(E') = 1 - p$. Observe that the event *some* $\mathbf{x}_i$ *is in* E is the same as the event *some* $\mathbf{x}_i$ *is not in* E'. Its complement is *each* $\mathbf{x}_i$ *is in* E', and this defines E_i'. Then,

$$\begin{aligned}\{\mathbf{x} \,|\, \text{each } \mathbf{x}_i \text{ is in } E'\} &= \{\mathbf{x} \,|\, \mathbf{x}_1 \in E'\} \text{ and } \cdots \text{ and } \{\mathbf{x} \,|\, \mathbf{x}_n \in E'\} \\ &= E_1' \cap \cdots \cap E_n'.\end{aligned}$$

By Example 21.1.3, $E_1', \ldots, E_n'$ is a list of pairwise independent events, so

$$\begin{aligned}\Pr(\text{each } \mathbf{x}_i \text{ is in } E') &= \Pr(E_1' \cap \cdots \cap E_n') \\ &= \underbrace{\Pr(E') \cdots \Pr(E')}_{n} \\ &= (1 - p)^n.\end{aligned}$$

Therefore, by the complement formula,

$$\begin{aligned}\Pr(\text{some } \mathbf{x}_i \text{ is in } E) &= 1 - \Pr(\text{each } \mathbf{x}_i \text{ is in } E') \\ &= 1 - (1-p)^n\end{aligned}$$

Here are a couple of examples of how to use the formulas in Examples 21.1.4 and 21.1.5.

EXAMPLE 21.1.6 Toss a die 25 times. The sample space of this experiment is

$$SS_{25} = \{(x_1, \ldots, x_{25}) \mid x_i \in \{1,2,3,4,5,6\}\}.$$

Choose an $(x_1, \ldots, x_{25}) \in SS_{25}$ at random. Find the probability that some $x_i \neq 3$.

Solution: The event $E = \{3\}$ has probability $\Pr(E) = \frac{1}{6}$. We are asked to find the probability that *some* $x_i \notin E$, or equivalently that *some* $x_i \in E'$. Then by Example 21.1.4,

$$\Pr(\text{some } x_i \neq 3) = 1 - \frac{1}{6^{25}}$$

EXAMPLE 21.1.7 Draw a card with replacement 7 times. The sample space of this experiment is

$$SS_7 = \{(x_1, \ldots, x_7) \mid x_i \text{ is a card}\}.$$

Choose an $(x_1, \ldots, x_7) \in SS_7$ at random. Find the probability that some x_i is a K or a Q.

Solution: The event $E =$ the set of K's and Q's in the standard deck. E has probability $\Pr(E) = \frac{8}{52}$. We are asked to find the probability that *some* x_i *is in* E. Then by Example 21.1.5,

$$\begin{aligned}\Pr(\text{some } x_i \in E) &= 1 - \left(1 - \frac{8}{52}\right)^7 \\ &= 1 - \left(\frac{44}{52}\right)^7\end{aligned}$$

21.2 Exercises

1. Let Exp be the experiment of tossing a die once. Repeat the experiment Exp sequentially 5 times, call the sequential experiment Exp_5 and let SS_5 be its sample space. Choose a sequence $(x_1, \ldots, x_5) \in SS_5$. Find the probability that some x_i is even. Answer: $E = \{2, 4, 6\}$, $\Pr(E) = \frac{1}{2}$. $\Pr(\text{some } x_i \text{ is even}) = 1 - \frac{1}{2^5}$.

2. Let Exp be the experiment of tossing two dice once. Repeat the experiment Exp sequentially 10 times, call the sequential experiment Exp_{10} and let SS_{10} be its sample space. Choose a sequence $((x_1, y_1), \ldots, (x_{10}, y_{10})) \in SS_{10}$. Find the probability that some sum $x_i + y_i$ is 7. Answer: $E =$ the sum of the faces is a 7. $\Pr(E) = \frac{1}{6}$, $\Pr(\text{some sum is 7}) = 1 - \frac{5^{10}}{6^{10}}$.

3. Let Exp be the experiment of drawing a card and replacing it. Repeat the experiment Exp sequentially m times, call the sequential experiment Exp_m and let SS_m be its sample space. Choose a sequence $(x_1, \ldots, x_m) \in SS_m$. Find the probability that some x_i is a Q. Answer: $E =$ choose a Q, $\Pr(E) = \frac{4}{52}$, $\Pr(\text{some } x_i \text{ is a } Q) = 1 - \left(\frac{48}{52}\right)^{10}$.

4. Let Exp be the experiment of drawing a card, replacing it, and then drawing another card. Repeat the experiment Exp sequentially t times, call the sequential experiment Exp_t and let SS_t be its sample space. Choose $((x_1, y_1), \ldots, (x_t, y_t)) \in SS_t$. Find the probability that some (x_i, y_i) is a pair (K, Q). Answer: $E = \{(x, y) \mid x \text{ is a } K \text{ and } y \text{ is a } Q\}$, $\Pr(E) = \frac{4^2}{52^2}$, $\Pr(\text{some } (x_i, y_i) \text{ is a } (K, Q) \text{ pair}) = 1 - \left(1 - \frac{4^2}{52^2}\right)^t$.

Chapter 22

Conditional Probability

If we are given events E and F then the symbol $E|F$ means E *given* F. This indicates that if F is a given event, then when examining E you must somehow observe only the outcomes of E that occur in F. Most importantly, *you must know that your experience is in* F. We define *the conditional probability of* E *given* F, written as $\Pr(E|F)$, as

$$\Pr(E|F) = \frac{\Pr(E \cap F)}{\Pr(F)} \text{ when } \Pr(F) \neq 0.$$

This formula *defines* the term *conditional probability.* The idea is that once F has occurred then $\Pr(E|F)$ is calculated using F as a sample space for a new experiment and $E \cap F$ as the event. Thus, we should think of $\Pr(E|F)$ as the probability of $E \cap F$ in the sample space F.

22.1 What Does Conditional Mean?

In this context, $\Pr(E \cap F)$ takes on a new meaning. Suppose that E and F are independent events. Then by the definition of independent events and the definition of conditional probability

$$\Pr(E|F) = \frac{\Pr(E \cap F)}{\Pr(F)} = \frac{\Pr(E)\Pr(F)}{\Pr(F)} = \Pr(E).$$

This shows us that

$\Pr(E|F) = \Pr(E)$ when E and F are independent events.

Thus, if E and F are independent events then E and $E|F$ have the same probabilities. Intuitively, if E and F are independent events then the probability of E does not depend on F at all. But we should not use our intuition to calculate $\Pr(E|F)$ because that can lead to mathematical problems.

If the outcomes in S are equally likely, then

$$\Pr(E \cap F) = \frac{n(E \cap F)}{n(\mathrm{S})} \text{ and } \Pr(F) = \frac{n(F)}{n(\mathrm{S})}.$$

Thus, in the case of equally likely outcomes, we can calculate conditional probability using the simplification

$$\Pr(E|F) = \frac{\Pr(E \cap F)}{\Pr(F)} = \frac{n(E \cap F)/n(\mathrm{S})}{n(F)/n(\mathrm{S})} = \frac{n(E \cap F)}{n(F)}.$$

We have thus found the following formula.

If the outcomes in S are equally likely then

$$\Pr(E|F) = \frac{n(E \cap F)}{n(F)}.$$

EXAMPLE 22.1.1 Experiment: Choose and remove a card from a standard deck, and then choose a second card. The sample space for this experiment is

$$\mathrm{S} = \{(A, B) \,|\, A \text{ and } B \text{ are cards from the standard deck and } B \neq A\}.$$

We have counted this kind of set before. $n(\mathrm{S}) = 52 \cdot 51$ since the second place B is different from the first place A.

We stick to our stated formulas. Let Kh and Qh be cards of the heart suit. Let $E = \{(Kh, X) \,|\, X \text{ is a card } Kh \neq X\}$, and let $F = \{(X, Qh) \,|\, X \text{ is a card } X \neq Qh\}$. Evidently $n(F) = n(E) = 51$.

1. Notice that $\Pr(E) = \dfrac{n(E)}{n(\mathrm{S})} = \dfrac{51}{52 \cdot 51} = \dfrac{1}{52}$.

2. Notice that $E \cap F = \{(Kh, Qh)\}$ so that $n(E \cap F) = 1$. Hence,

$$\Pr(E|F) = \frac{n(E \cap F)}{n(F)} = \frac{1}{51} \neq \frac{1}{52} = \Pr(E)$$

Thus, E and F are not independent events.

This is not in conflict with Example 20.1.3, since in Example 20.1.3 we require that we work with replacement when defining the experiment with sample space $\mathrm{S} \times \mathrm{S}$.

EXAMPLE 22.1.2 There is a bag of black and white chips. The black chips are numbered 1 to 13 and the white chips are numbered 1 to 13. Experiment: Choose a chip, observe its color and number, remove it from the bag, and then choose another chip, observing its color and number. The sample space for this experiment is

$$\mathrm{S} = \{(x, y) \mid x, y \text{ are chips in the bag}, x \neq y\}.$$

Then $n(\mathrm{S}) = 26 \cdot 25$. We draw a tree diagram to analyze the problem of drawing a black or a white chip.

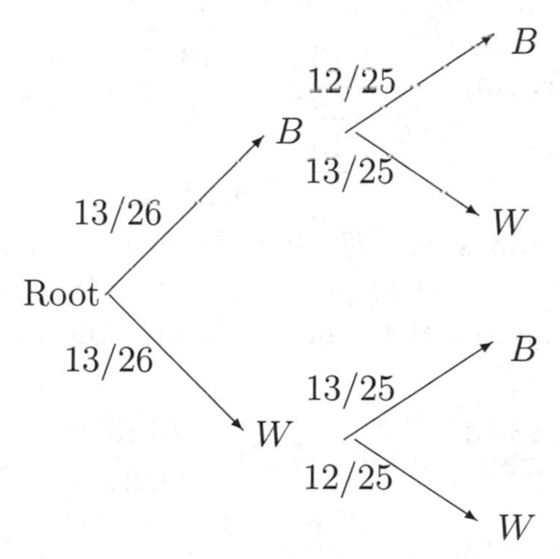

There are 13 black chips and 26 chips total, so the probability of choosing a black on the first choice is 13/26. This is also the probability for choosing a white chip on the first choice of chips. Having chosen a black chip there are 12 black chips, 13 white chips, and 25 chips left in the bag. Thus, the second black chip is chosen with probability 12/25, and the white is chosen on the second choice with probability 13/25. Similar analysis applies to the lower branches of the tree.

From our tree diagram we can calculate some important conditional probabilities associated with this problem.

1. The probability of *choosing at least 1 black chip* in this experiment is found by first finding the paths on the tree diagram beginning at the Root and that contain at least 1 B. Then calculate the probabilities corresponding to these paths using the multiplication principle, and then add these probabilities together using the inclusion/exclusion principle. I will give more detail.

The paths from the Root that contain a B correspond to a choice of a black chip. There are 3 such paths. They contain the vertices BB, or BW, or WB. These are the choices we wish to count. The two edges in the path correspond to two events done in sequence, so we use the multiplication principle and multiply the probabilities labeling the edges in this path. The probabilities of the paths are

$$\Pr(BB) = \frac{13}{26}\frac{12}{25}, \ \Pr(BW) = \frac{13}{26}\frac{13}{25}, \text{ and } \Pr(WB) = \frac{13}{26}\frac{13}{25}.$$

Thus, the probability of choosing at least 1 black chip is

$$\begin{aligned}\Pr(\text{at least 1 black chip}) &= \frac{13}{26}\frac{12}{25} + \frac{13}{26}\frac{13}{25} + \frac{13}{26}\frac{13}{25} \\ &= \frac{13}{26} + \frac{13}{26}\frac{13}{25}\end{aligned}$$

2. The probability of *choosing a black and a white chip* is found by finding the paths on the tree diagram that contain both a B and a W vertex. Those paths are BW and WB, and the associated probabilities are

$$\Pr(BW) = \frac{13}{26}\frac{13}{25} \text{ and } \Pr(WB) = \frac{13}{26}\frac{13}{25}.$$

Thus,

$$\begin{aligned}\Pr(\text{a black and a white}) &= \frac{13}{26}\frac{13}{25} + \frac{13}{26}\frac{13}{25} \\ &= 2\frac{13^2}{26 \cdot 25}.\end{aligned}$$

A note about the language used here. The problem asks that we choose a black and white chip. No order of these choices is given. Thus, we need to count choices of chips that contain a black chip or a white chip. The order in which they come to us is not specified, so we will count the event where we first choose a black chip followed by a choice of a white chip, and the event where we first choose a white chip followed by a choice of a black chip. These events correspond to the 2 paths on the tree as indicated above.

3. The probability of *choosing a black chip on the second draw* is found by finding the paths in the tree that begin with the Root and end in a vertex labeled with B. There are 2 such paths and they contain BB and WB. Then

$$\begin{aligned}\Pr(\text{a black on the second draw}) &= \frac{13}{26}\frac{12}{25} + \frac{13}{26}\frac{13}{25} \\ &= \frac{13}{26}\end{aligned}$$

EXAMPLE 22.1.3 Use the bag defined in the above example. Choose a chip, observe its color and number, remove it from the bag, and then choose another chip, observing its color and number. As above, $n(\mathrm{S}) = 26 \cdot 25$.

We draw a tree diagram to analyze the problem of drawing a chip numbered 1.

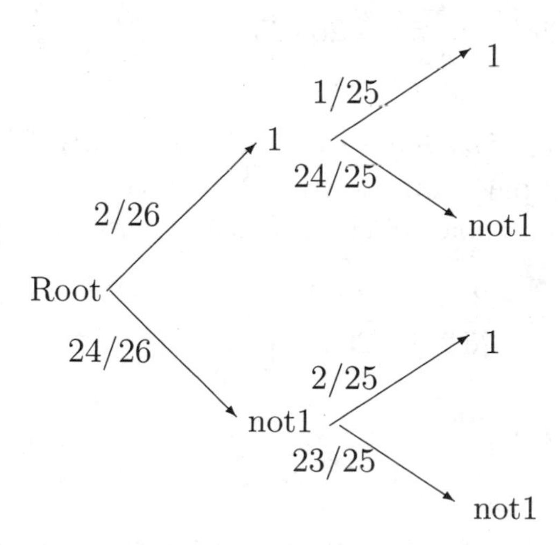

Let me explain how the tree is labeled. There are 2 chips numbered 1, so the probability of choosing a 1 on the first choice is 2/26. The probability for choosing a chip not numbered with 1 is 24/26 on the first choice of chips. Thus, the second chip numbered with 1 is chosen with probability 1/25, and the second chip chosen that is not numbered 1 is 24/25. Similar analysis applies to the lower branches of the tree.

From this diagram we can calculate some important conditional probabilities associated with this problem.

1. The probability of choosing *at least 1 chip numbered 1* is found by finding all of the paths from the Root that contain a vertex labeled 1. There are 3 such paths. Thus,

$$\begin{aligned}\Pr(\text{at least 1 chip numbered 1}) &= \frac{2}{26}\frac{1}{25} + \frac{2}{26}\frac{24}{25} + \frac{24}{26}\frac{2}{25} \\ &= \frac{2}{26} + \frac{24}{26}\frac{2}{25}\end{aligned}$$

2. The probability of choosing *a chip numbered 1 and a chip not numbered 1* corresponds to 2 paths. The associated probability for this event is

$$\frac{2}{26}\frac{24}{25} + \frac{24}{26}\frac{2}{25} = \frac{2 \cdot 2 \cdot 24}{26 \cdot 25}$$

3. The probability of *choosing a chip numbered 1 on the second choice* corresponds to 2 paths on the tree. Notice that we make no conditions about the first choice of chip. The associated probability for this event is

$$\frac{2}{26}\frac{1}{25} + \frac{24}{26}\frac{2}{25}.$$

22.2 Exercises

1. Draw a card from a standard deck, note its kind, remove it, and choose another card. Let E be the event that the first card was

a K, (a King), and let F be the event that the second card is a Q, (a Queen). Draw and label the associated tree diagram.

2. Draw a card from a standard deck, note its kind, and remove it from the deck. Then draw a second card and note its suit. Draw and label the associated tree diagram for this experiment.

3. In the previous problem, determine if the event of choosing a K on the first draw is independent of the event of choosing a Q on the second draw.

4. There is a bag containing black chips numbered 1 to 10, and white chips numbered 1 to 10. You choose one chip and remove it, noting its color. Then you choose another chip and note its number. Draw and label the associated tree diagram.

Chapter 23

Bayes' Theorem

The purpose behind the examples in the previous chapter is to show you that some probabilities can be viewed geometrically and more clearly on a tree. The branches give you almost a mechanical method for finding probabilities that otherwise require some focused mathematical thought. In this section we use a tree diagram to make an abstraction of the solution method used in those examples.

23.1 The Theorem

Suppose you are given an event E and that you wish to study its dependence on an event E. The formula for studying $\Pr(E|F)$ that we are interested in is called *Bayes Theorem*. We will derive it and then relate it to a tree diagram.

By the definition of conditional probability,

$$\Pr(E|F) = \frac{\Pr(E \cap F)}{\Pr(F)}. \tag{23.1}$$

Here is where we do a clever thing. We do it because it works, because some genius had the insight to do this first. We follow along, thinking like this genius.

$$\Pr(F|E) = \frac{\Pr(E \cap F)}{\Pr(E)}$$

so

$$\Pr(E)\Pr(F|E) \quad = \quad \Pr(E \cap F). \tag{23.2}$$

The set F can be written as a disjoint union

$$F = (E' \cap F) \cup (E' \cap F).$$

This simply means that something in F is in E or it is in E', but it is not in both. Since

$$(E \cap F) \cap (E' \cap F) = (E \cap E') \cap F = \emptyset,$$

$(E \cap F)$ and $(E' \cap F)$ are mutually exclusive events, which means that

$$\begin{aligned} \Pr(F) &= \Pr((E \cap F) \cup (E' \cap F)) \\ &= \Pr(E \cap F) + \Pr(E' \cap F). \end{aligned} \tag{23.3}$$

Substituting equation (23.2) into equation (23.3) yields

$$\begin{aligned} \Pr(F) &= \Pr(E \cap F) + \Pr(E' \cap F) \\ &= \Pr(E)\Pr(F|E) + \Pr(E')\Pr(F|E'). \end{aligned} \tag{23.4}$$

(The second substitution uses E' in place of E.) Finally, combining equations (23.1), (23.2), and (23.4) brings us to the formula we seek.

$$\begin{aligned} \Pr(E|F) &\overset{(23.1)}{=\!=} \frac{\Pr(E \cap F)}{\Pr(F)} \\ &\overset{(23.2)}{=\!=} \frac{\Pr(E)\Pr(F|E)}{\Pr(F)} \\ &\overset{(23.4)}{=\!=} \frac{\Pr(E)\Pr(F|E)}{\Pr(E)\Pr(F|E) + \Pr(E')\Pr(F|E')} \end{aligned}$$

BAYES' THEOREM 23.1.1 *Let E and F be events in a sample space* S. *Then*

$$\boxed{\Pr(E|F) = \frac{\Pr(E)\Pr(F|E)}{\Pr(E)\Pr(F|E) + \Pr(E')\Pr(F|E')}}$$

A paraphrasing of this theorem is that we can find $\Pr(E|F)$ if we know $\Pr(F|E)$ and $\Pr(F|E')$.

There is no shortcut to conditional probability in Bayes' Theorem, but a tree diagram can help. Suppose you wish to examine $\Pr(E|F)$ along the lines of Bayes' Theorem. You might construct a tree like the one below.

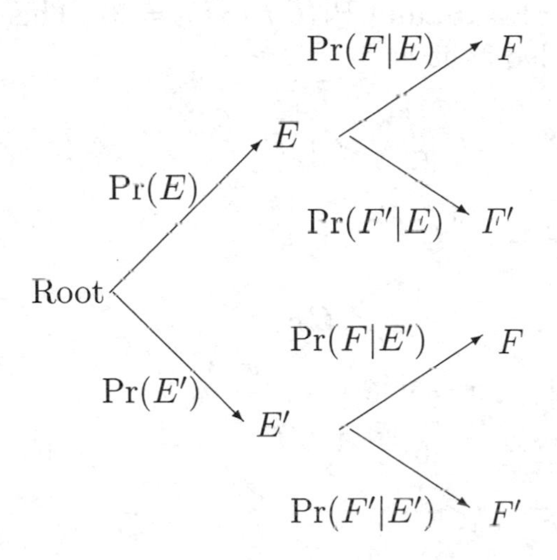

This tree gives us more information about the probabilities in this problem than we are looking for, but it still contains all of the information we need to find $\Pr(E|F)$ in terms of other supposedly simpler probabilities. Intuitively, then, Bayes' Theorem shows that $\Pr(E|F)$ is given by finding the probabilities along the path from Root to E to F, and then dividing by all paths that end in F. The problem of finding $\Pr(E|F)$ given a tree diagram like the one above is just substitution now.

EXAMPLE 23.1.2 A certain computer graded exam is given a grade of pass/fail. We note that 75% of the students passed the test on their own merits (a student pass), while 25% failed the exam. However, technology fails once again and the computer assigned a passing grade to 60% of the tests (computer passes) and a failing grade to the remaining 40%. The manner of assignment makes the event of a computer pass independent of the event of a student pass. Find the probability that a student passed the test given that they received a computer pass.

Solution: Let SP be the set of students who passed, and let SF be the complement of SP which is the set of students who failed. Let CP be the set of computer passed students, and let CF be the complement of CP, which is the set of students failed by the computer. Of course, $\Pr(SP) = .75$ and $\Pr(CF) = .25$. The computer assigned grades had nothing to do with the actual grade, so $\Pr(CP|SP) = \Pr(CP) = .6$, and $\Pr(CP|SF) = .6$. This is why the tree diagram is labeled as it is.

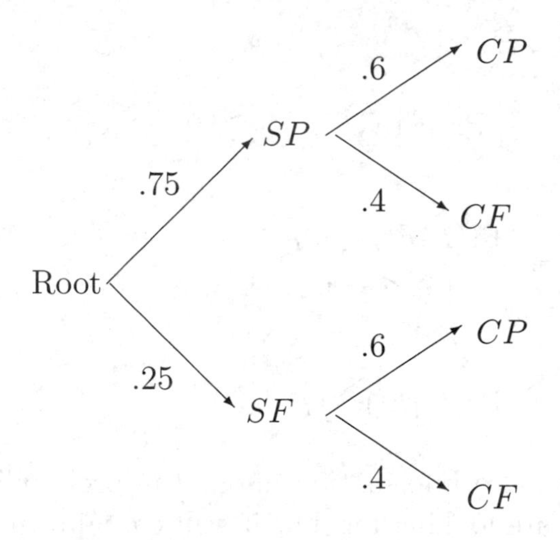

We are asked to find $\Pr(SP|CP)$. We use the tree diagram and Bayes' Theorem. By Bayes' Theorem

$$\begin{aligned}\Pr(SP|CP) &= \frac{\Pr(SP)\Pr(CP|SP)}{\Pr(SP)\Pr(CP|SP)+\Pr(SF)\Pr(CP|SF)} \\ &= \frac{(.75)(.6)}{(.75)(.6)+(.25)(.6)} \\ &= .75\end{aligned}$$

The reason that there is so much cancellation in the calculation of such an intuitively clear problem is that SP and CP are independent probabilities. We could have solved the problem by just observing that $\Pr(SP|CP) = \Pr(SP) = .75$. Think of this example as a first exercise in how to use Bayes' Theorem, and Don't look for a deep underlying meaning behind it.

EXAMPLE 23.1.3 This holiday season about 25% of the people on the road will be impaired. A new test for impairment is accurate about 99% of the time. Find the probability that a randomly chosen driver escapes arrest for impaired driving. That is, find the probability that he is not impaired given that his test is negative.

Solution: Let M be the set of people who are impaired, let P denote the event that a positive test for impairment is recorded, and N denotes a negative result. If you are impaired, then $\Pr(P|M) = .99$ because the test is 99% accurate. If you are not impaired M' then $\Pr(P|M') = .01$ for the same reason. That is why the tree diagram is labeled as it is.

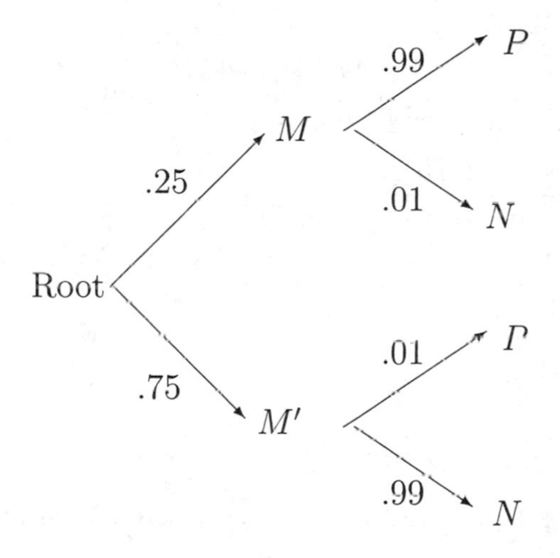

We are asked to find $\Pr(M'|N)$. We use the tree diagram and Bayes' Theorem. By Bayes' Theorem we have

$$\Pr(M'|N) = \frac{\Pr(M')\Pr(N|M')}{\Pr(M')\Pr(N|M') + \Pr(M)\Pr(N|M)}$$

and by the tree diagram labels we have

$$\begin{aligned}\Pr(M'|N) &= \frac{(.75)(.99)}{(.75)(.99) + (.25)(.01)} \\ &= \frac{99}{99.22\cdots}.\end{aligned}$$

So almost 100% of the population of people who received a negative test result are also unimpaired.

EXAMPLE 23.1.4 Use the same test, the same data, and the same social setting given in the above example.

1. It is interesting to note that $\Pr(N|M) = (.25)(.01) = .0025$, so only .25% of the impaired population will test negative.

2. Find the probability that you received a *false positive test.* That is, you are not impaired, but you tested positive. You are out to calculate $\Pr(M' \cap P)$, which is just $\Pr(M')\Pr(P|M') = (.75)(.01) = .0075$. So a scant .75% of the population will receive a false positive test.

3. What percentage of the people test positive and are unimpaired? We are looking for $\Pr(P \cap M') = \Pr(P)\Pr(M'|P)$. By Bayes' Theorem,

$$\begin{aligned}\Pr(M'|P) &= \frac{\Pr(M')\Pr(P|M')}{\Pr(M')\Pr(P|M') + \Pr(M)\Pr(P|M)} \\ &= \frac{(.75)(.01)}{(.75)(.01) + (.25)(.99)} \\ &= \frac{1}{34}\end{aligned}$$

which is less than 3%. Hence,

$$\Pr(P \cap M') \leq (.01)(.03) = .0003.$$

Thus, less than .03% of the people will test positive and will be unimpaired.

5. Because $\Pr(E|F)$ requires that you know you are in the event F, and since *many impaired people do not know that they are impaired*, $\Pr(E|F)$ would not be the probability to best describe the end result of this experiment. The intersection seems to be a better measure of a false positive.

23.2 Exercises

Refer to exercises 1 through 3. Let us say that about 33% of people have red hair. A new test for the red hair gene is accurate about 90% of the time.

1. Draw the associated tree diagram.

2. Find the percentage of people who received a false positive test. That is, they do not have red hair, but they tested positive for the gene. Answer: $(1 - .33)(.1)$ since the test is inaccurate.

3. Find the probability that a randomly chosen person has red hair given that they tested positive for the red hair gene. Answer: $(.33)(.9)$

Draw a card from a standard deck, remove it, and then choose another card. Make note of your choices. K denotes a King and Q denotes a Queen.

1. Draw the associated tree diagram.

2. Find the probability that you chose a K and another kind of card. Answer: $\frac{4}{52}$.

3. Find the probability that your second card is a K given that you chose a K as your first card. Answer: $\frac{4}{52}\frac{3}{51}$

4. Use Bayes' Theorem to find the probability that your first card is a K given that you chose a Q as your second card. Answer: $\frac{1}{22}$

Chapter 24

Statistics

24.1 Introduction

In each text on finite mathematics there is a section or two on statistics. The student is guided through the calculation of median and mean, and perhaps standard deviation. We will break with tradition in this chapter and guide the student past some common inaccuracies that seem to be pitfalls for those using or studying statistics. Specifically, we will examine the correct use and meaning of the ideas probability and statistics, median, and random samples.

Throughout this chapter, let Exp be an *experiment*, and let S be its *sample space*. When Exp is carried out many times we call the collected outcomes a *survey*.

Let X be the set of pairs (x, n) where $x \in \mathrm{S}$ is one of the observed outcomes of the experiment, and where n is the *frequency of x in the survey*, or equivalently, n is the number of times that x occurs in the survey. We will call X the *data of the survey*. The *frequency graph* $\Gamma(X)$ of X, or *the frequency graph of the survey*, is the bar graph produced by graphing the pairs $(x, n) \in X$ in the plane.

24.2 Probability Is Not Statistics

When dealing with probability, we must be careful not to make an error that is common outside of the classroom. Some are using a probability as a statistic. While each statistic comes from a probability, many probabilities are not statistics. A probability is not a future predicting statistic until it is associated with many trials of the associated experiment Exp that defines it.

While a probability will tell you the percentage of time that you will choose an outcome from a specific event from a sample space, if the choices are made randomly over an infinite number of trials, that probability cannot be used to determine the percentage of times that a future bit a data will be in that event. In other words, probabilities measure what has happened in an experiment while a statistic with enough data points can be used to predict future behavior in an experiment to within a small error. The two should not be confused.

24.3 Conversational Probability

In the mid-seventeenth century there were two scholars of a wide-ranging intellect. They thought about and published their work in algebra, calculus, physics, philosophy, the law, theology, and most of all in probability. Indeed, because of the letters exchanged, we call these two mathematical giants the founders of present day probability. We will eavesdrop on these two as if they were talking in their kitchen about inaccurate mathematical thinking of the early twenty-first century.

Pascal: Good day, good sir. Sharpen your wit my friend, for today I have a series of problems in the theory of probability that require our attention. These are not the problems that we considered, those of money, expected return on gambling investment, and on better strategies for gaming purposes. These are problems that show that during the 450 years since we first studied these problems, mankind has gotten sloppy in how it reads and writes about these ideas.

Fermat: I stand ready for your intellectual barrage, but let me ask. Why should we care? We founded this area of philosophical thought. We introduced the notions that modern man applies to his gambling. He even uses some methods in predicting the future, and he is good at it in some regards. But I agree. Since our time, man has become sloppy in his thinking about the elegant subjects of probability and statistics. Tell me your ideas.

Pascal: Men have preserved the fact that in order to measure probability, one must first define, in the best spirit of the scientific method, the experiment that is being executed. Once that experiment has been stated and observed, the experimenter must make a list of all of the outcomes of said experiment. The collection of all such outcomes is called the sample space. One cannot know what probabilities will appear until they know with certainty the experiment and the sample space.

Fermat: I see, my colleague. So if the experiment is to flip a coin once, then the sample space would be the set {Head, Tail}. These are all of the possible outcomes for this experiment. Or you might toss a spotted six sided die. The outcomes in this experiment are one of the six sides, so the sample space is $\{1, 2, 3, 4, 5, 6\}$. If we choose a card from the standard deck of 52 cards then the outcomes are any of the 52 cards, so the sample space is the standard deck of 52 cards. Things, I suppose, are more complicated if in your experiment you take 2 cards from the deck.

Pascal: Yes, my friend, and when speaking of that many outcomes, a description is more lingual and not often a list. Also, it is advantageous to simply count the number of outcomes in the sample space.

Fermat: So one counts $51 \cdot 26$ ways to choose 2 cards from 52, and one counts 11 outcomes from tossing two dice.

Pascal: Although your first guess is correct, your second reveals one of the intellectual obstacles confronting modern practitioners of practical probability. If we wish to write the most expressive sample space describing the outcomes of tossing a pair of dice and adding the two revealed sides, then our sample space must reflect the fact that 2 can appear only one way while 7 can appear 6 different ways. Here is how I think we can write a more expressive sample space.

Suppose the two dice are colored. One is white and the other is black. Then when listing the outcomes in our experiment, we would list a white 1 and a black 1 as (1,1). We would list a white 3 and a black 4 as (3,4) while we would list a white 4 and a black

3 as (4,3). With this distinction, a white number being first and a black number being second, we have exactly 6 ways for a 7 to occur. Namely, (1,6), (2,5), (3,4), (4,3), (5,2), and (6,1).

Fermat: I say the game is played with ivory dice, or red dice, but not white and black dice. Your game with different colored dice is certainly different from my game of craps.

Pascal: No, not really. The outcomes should be the same and come in the same frequencies no matter who plays the game. Thus, if the player is blind, he will play the game as though the dice were the same color. He simply feels the recessed dots on the top of the dice, and so he experiences the game differently than the sighted man. However, the number of ways to roll a 7 should be the same for him as it is for you. Your blind friend would have to assume the same number of ways to roll a 7 as you did when you rolled those white and black dice.

Fermat: God Bless Diophantus. You mean the probability of a 7 changes for me when I am old and blind?

Pascal: No. I mean the opposite of that. The probability of tossing dice does not depend upon its color combination or your eyesight. The outcomes from tossing a pair of dice are independent of the visual displays used on the dice, and the state of your corneas and retinas.

Fermat: I see. You are saying that if I gamble thinking that 7 occurs once every 11 tosses, then I will loose a great deal of money. But if I bet assuming that 7 occurs 1 times in every 6 tosses then I am more likely to earn those ill gotten gains.

Pascal: You have uncovered the underlying principle. The difficulty in using the numbers 2 through 12 as a sample space is that they are not equally likely to occur. Any computer in these times can be used to show that 7 will occur far more often than a 2 when fair dice are tossed.

Fermat: Let me tell you of an interesting subject discovered since the sixteenth century. When you or I want to know when the entire French people will buy their bread, we would have to ask all of them. By the time you had asked that last Frenchmen, we would be asking about when he *bought* his bread and not about when he *will buy* his bread that week.

With the use of their computers, todays political leaders take what is called a poll of 1000 people, which will tell them about more than the 1000 people polled. Let us say this survey shows that 200 out of 1000 will buy bread on Monday. That is 20%.

Pascal: Who cares about 1000. You said the problem was to know about *all* Frenchmen. You cannot know what the 1001-st Frenchmen will do based on a knowledge of the 1000. You must somehow include that next Frenchmen in your survey.

Fermat: That is correct, as far as I have told you. Given the information you have, you cannot tell what the next Frenchmen will do. You cannot say, for instance, that 20% of all Frenchmen will buy bread on Monday because 20% of the *surveyed* people bought bread on that day. Something more must be introduced.

Pascal: You mean that if I know more about these 1000 people then I can say that 20% of the French people will buy bread on Monday? That is amazing. It must be magic. Or maybe its from a civilization we never knew. What more can be said about this survey sample?

Fermat: It is called *randomness*. If we can ensure that as a group the 1000 people possess what is called randomness then we can use the 1000 to extrapolate what the entire nation will do. Well, at least up to a couple of percentage points. So, if we are careful, then we can say that the percentage of the population that will buy bread on Monday is between 18% and 22%. That is a good answer given what we are asking these 1000 people to do. Given randomness in our 1000 surveyed people, we can imply what hundreds of millions of people will be doing.

Pascal: This is remarkable. How do we go about finding a survey sample of 1000 people that possesses randomness? Is it difficult? Is it expensive?

Fermat: You can find a random sample of 1000 people, but it costs a great deal of money. The population has to be found well ahead of the execution of the experiment of buying bread on Monday. The ages, the races, income brackets, property ownership, and the questions being asked are just a few of the factors that have to be considered when choosing a random sample from a larger population. But there are those in government whose expertise is to collect survey samples that possess randomness.

Pascal: So a government can tell from a random sample of 1000 people how the entire population of the nation thinks or feels about controversial issues. Businesses must use this method to find how many people will buy their product.

Fermat: Indeed they do, and with good effect. Now, let me introduce a sour note. Some there are who think that they can without effort produce a survey sample possessing randomness by hiring people to ask questions of other people.

Pascal: Don't they know that one person will choose another person based on how their interlocking neuroses fit? Unless the person taking the survey is extensively schooled on choosing a sample, there is little hope that the total survey sample will possess randomness. *People choosing people will not result in a random sample* for your experiment.

Fermat: Good point. Let me expand on that for you. Suppose that you wish to know how many Frenchmen will buy their bread on Monday. A sample of any 1000 people will not make a good sample for your survey. For example, suppose you choose your survey on Monday at the bakery. Your survey would conclude that all Frenchmen buy bread on Monday, a totally uncalled for deduction.

Pascal: So 1000 is not enough for randomness. How many do you need?

Fermat: You do not need more people in your survey. You need randomness. These are two different things. Here is an example. You decide you need more people. You go to America and survey all of the American people. That is a survey of several hundreds of millions of people. You discover in your survey that the American people will buy their bread on Saturday. From the sheer size of the survey population you conclude that your survey sample possesses randomness. Thus, you conclude, all Frenchmen will buy their bread on Saturday, an utter falsehood.

Pascal: So where did I make my error?

Fermat: The error is to think that more is more accurate when taking surveys. This is wrong, for what has the American public got to do with the life of a Frenchmen?

Pascal: I see. More did not make it random, did it? I can see that. But who would ask the Americans about the buying habits of the French?

Fermat: This is the kind of error made by some who have set themselves up as professional pollsters or professional survey takers. Those who have not invested enough money into the construction of their survey, or whose population data bank is not extensive will inevitably mislead their customers, for example, by telling them that one group measures the habits of another if the size is large enough. This is more common than people, newspaper editors, and politicians want to believe.

Pascal: Are there other errors in assuming randomness?

Fermat: Yes. The most common is in allowing people to choose people, and then to call it a random choice. Randomness does not happen in this way, no matter how loudly the people doing it yell. In fact, the louder they yell, the more sure you can be that their survey sample is anything but random.

Pascal: Let me take the lead in our next topic.

Fermat: You may, my friend, as long as I can urge you on.

Pascal: My digression deals with the present generation's confusion between probability and statistics.

Fermat: You must be confused yourself, for isn't each statistic a probability?

Pascal: Yes. Each statistic is defined by a probability. But that is where the similarity changes. Because you see, not every probability is a statistic.

Fermat: That subtlety requires a detailed explanation, sir. Your claim is that every statistic is defined by a probability, but most probabilities are not statistics. You are required to tell me how that is possible.

Pascal: I will give a few examples to explain. Firstly, a statistic is given by a probability. Suppose you know a batter named D in the American game of baseball who for the first 27 weeks of the season has a batting average of .333. This means that the number of hits made by D during that time when divided by the number of at bats during that time is .333.

Put another way, in the first 27 weeks of the season, he gets a hit about one third of the time. If you randomly choose an at bat from these 27 weeks of at bats, and if you make this choice over many times, then about 1/3 of the time you will choose an at bat in which he hits the ball.

Fermat: So if I am reading about D's at bats during this 27 week period, and if I read these statistics at random, then after many readings I will notice that 1/3 of the readings produced a hit. Then if I choose any 12 at bats from the period, I can say that 4 of the 12 of D's at bats will be a hit. Correct?

Pascal: No, and that is the confusion I am talking about. It is possible that a batting average will be .333 but that a given series of 12 at bats will feature no hits. This is called a hitting slump, and it occurs quite often in the game. And this is the confusion.

The number .333 simply means that over many choices of at bats, you will read about a hit in about 1/3 of the time. To claim anymore of the number .333 is a misuse of the average. The average you see is a probability, while the predictive use of a number treats it like a statistic. Probabilities do not predict outcomes, except when many outcomes are considered in calculating the probability.

Fermat: Then if D's average is .333 over 27 weeks, just how many hits can I expect in a series of 12 at bats?

Pascal: You will find any number of hits during the 12 at bats you choose. In fact, our batter D batted .333 between April and the middle of September, and he went 0 for 12 after that. Thus, his batting average could not be used to determine how he would hit in that given series of at bats. In the future, D might get 12 hits in a row, or he might go hitless. The fact is that he might have any number of hits during those 12 at bats.

Fermat: Can an athlete do this? You would think that if he was so regular for that much of the season, then he would be regular in the 12 at bats. What would cause such a decline in his batting?

Pascal: The game changes from pitch to moment to day. With this kind of variability, it is likely that a few bad days for the batter will match up with a few good days for the opponents. But more likely is the fact that pressure affects the batter, and pressure changes from day to day. Maybe D was having a distracting 12 days with his girl friend. Or maybe there was another pressure within the game that distracted D. The point is that .333 does not measure these external parameters.

Fermat: So D's average does not tell us how he will do over a period of time within those 27 weeks. But how does it predict D's performance over future at bats?

Pascal: Well, we just decided he might go 0 for 12 after batting .333 so I suppose your question is about how he might perform in the next at bat. Just one more at bat than he has taken. I hope you will enjoy this little nugget.

Fermat: So let me set up the discussion, please. D has a batting average of .333 for the first 27 weeks of the season. He steps to the plate to begin the 28-th week. I guess that D's average on the next at bat will be .333.

Pascal: This is again a confusion practiced by many a modern baseball fan. If you think about it for a moment, you cannot have a fractional at bat. You either hit the ball or you miss it. Once done, that first at bat in week 28 is over. You do not get to replay it again. D's average for that one at bat is 0 or 1, and nothing else.

Fermat: So once again the .333 batting average does not affect D's chances of getting a hit.

Pascal: No. I did not say that. I said that your average in that *one* at bat is either 0 or 1. And you cannot determine that one at bat average until after you take that one at bat. The .333 might tell you what kind of person D is at the plate, but it will not tell you how D will perform at the plate that one time.

Fermat: Then how can I incorporate that one at bat in .333?

Pascal: Let us make an example of it. D has a .333 batting average, and then he takes an at bat. Let us say that D hits the ball. Then, if the numbers are good, and in many cases they are, D's average over the 27 weeks *and one at bat* is now .334. By getting a hit in that one time at the plate, D changed his seasons batting average from .333 to .334. On the other hand, suppose that D does not hit the ball in that one at bat. Then D will watch his season's batting average go from .333 to .332. Missing the ball in one at bat lowered D's average a little.

Fermat: But why can't I use the .333 to predict that one at bat?

Pascal: *Because you have not said what you wish to predict.*

What will be the conclusion of your guess? Will you say that .333 of the time D will get a hit? That is wrong, because there is only one at bat that you are considering. Do you mean by .333 that all things being equal D will hit the ball 1/3 of the time. That is wrong, because in a good game, and baseball is a good game, things are never the same two times in a career, much less in a row. Even the pitches thrown will change from pitch to pitch and from day to day. Expecting to use .333 to predict that at that one at bat D will hit or go hitless is also wrong because on that one at bat D's average will be 0 or 1. But if I have runners in scoring position then D the man I want at the plate.

Fermat: You seem to have a love of the variability of the game. Your energy gives me the jitters.

Pascal: Allow me to end with a brainteaser. Your season's batting average is .333 through 27 weeks. While you cannot say with certainty what will happen to you on that next at bat, you might be able to make a statement about all of the players who ever played the game. The brainteaser is this. Suppose that there are 1000 players in the game of baseball who had a .333 batting average through the first 27 weeks of the season. What number of them got a hit on the next at bat? Is the answer 333?

Fermat: I would have to be careful on this one, because the temptation is to treat .333 as a statistic instead of as a probability.

Pascal: Shall we suspend our conversation?

Fermat: Of course. You'll want coffee and French onion soup, I suppose? I'll take a cup-a-tea.

24.4 Conditional Statistics

Suppose that E and F are events in a sample space S. Then the *conditional probability of E given F* is calculated as

$$\Pr(E|F) = \frac{\Pr(E \cap F)}{\Pr(F)}.$$

This is the probability of an outcome S being in the event E if you know that said outcome is in the event F. There is no problem with the calculation, as long as it is taken only as a calculation. When one begins to use $\Pr(E|F)$ as a predictive statistic, then there is a problem.

EXAMPLE 24.4.1 There are numerous internet businesses that encourage your business by arguing as follows. Let F be the event that you will hire this firm, and let E be the event that you make money over all. Suppose the firm advertises that in their advertisements that 60% of the people who hired them in 1999 ended up making money. In the more questionable ads, the firm implies either directly or indirectly that your chances of making money with them is then 60%. This uses $\Pr(E|F)$ as a statistic. It is not.

$\Pr(E|F)$ is a probability, and just like our baseball batting average, it cannot be used to predict how the next customer entering through the door will benefit. It can only be used to know what the previous customers have done. In floating a number like 60% to be absorbed by the innocent customer, the firm is taking advantage of the general population's cluelessness concerning the use and practice of probability in the market place.

Another example is the reliability of a test.

EXAMPLE 24.4.2 Let F be the event of pregnant woman *in the world*, and let E be the event that a pregnancy test yields a negative result. If a pregnant woman has a negative test result then we call the result a *false negative*. Many texts consider only the existence of a *false positive result*, but we feel there is more to learn with *false negative results*.

So you walk into the room, pregnant, and you take the pregnancy test. The result is a negative. If we suppose that $\Pr(F) = .30$ and that $\Pr(E \cap F) = .01$. Then $\Pr(E|F) = \dfrac{.01}{.30} = .0\overline{333}$, which means that if you are already pregnant (in event F), then the probability that your test result is negative is quite small. But we ask, *if you that you are pregnant then why are you taking the test?*

However, the number $.0\overline{333}$ applies to the very large population of pregnant females in the world. This probability does not help you if the event F consists of pregnant females in Seattle and you are testing in Dubuque. In that case, $.0\overline{333}$ does not refer to your probability of a false negative result. The population used for this statistic came from Seattle and you are not from there, so their statistics do not apply you. For what has a pregnant Seattle woman got to do with some Dubuque debutant's reproductive state?

But there is a more important question to be raised here.

EXAMPLE 24.4.3 Suppose that you walk into the testing facility not knowing or suspecting your reproductive state, and the test result is negative. Using $\Pr(F) = .30$ and that $\Pr(E \cap F) = .01$ then the probability that you have a false negative is $.0\overline{333}$, while the probability that you are pregnant and have a false negative is the substantially smaller probability .01. Which probability should be used when consulting with this woman? Which is a better measure of her physical experience? If we give her the wrong number then aren't we advising irresponsibly? No informed response has yet been offered.

Another example is the ambiguous drunk.

EXAMPLE 24.4.4 You step out of your car, not knowing your intoxication level at all. You are then asked to take the balloon test, which shows that you are over the limit. Which probability should we use? Which one accurately describes the situation that you are in. I suggest that $\Pr(E|F)$ is out of the question because its use means that we knew you were drunk to begin with. Since this is not the case, a better measure would be $\Pr(E \cap F)$ which gives the probability that you are drunk *and* that your test showed that you were over the limit. After all, your roll in this experiment was to step out of the car and then take a test. This seems to rule out using you as an outcome in F, the set of those intoxicated souls.

Therefore, the use of $\Pr(E|F)$ as a predictive value in the presence of the uninformed public can be questioned.

24.5 The Mean

There are a few numbers associated with statistical surveys that can be calculated from X and the frequency graph $\Gamma(X)$. A popular number is called *mean.* The mean has no meaning for X unless S is a set of numbers x. For example, you cannot calculate the average of a survey if each $x \in$ S is an eye color, or more generally if each $x \in$ S is other than a number.

EXAMPLE 24.5.1 Suppose you go out into the world and count the number of times you see one of the eight primary colors. Your sample space is S = the set of the eight primary colors, and your data set X is the set of pairs (*color*, n) where *color* is one of the eight primary colors, and where n is the number of times you saw *color.* You would not calculate the mean of such a survey as there is currently no scientific means to determine the average of the set of eight colors.

If S is a set of numbers then we can calculate the mean for the survey X, but this requires us to construct our survey cleverly, and with a good deal of insight into the problem being surveyed. With this in mind, in this chapter *we will restrict our attention to a limited kind of experiment whose outcomes are numbers.*

Let us move on to the calculations.

Let Exp be an experiment used to conduct a survey, let S be the sample space of Exp, and let X be the data of the survey. The mean is the *average* that is taught in high schools, and it is calculated as

$$\text{mean of } X = \frac{\sum\{nx \mid (x,n) \in X\}}{\sum\{n \mid (x,n) \in X\}}.$$

Since $(x, n) \in X$ means that x was found a total of n times in the survey, then $\sum\{nx \mid (x,n) \in X\}$ is the sum of the numbers x that were found in the survey, *including the number of times that the number x occurred in the survey.* One cannot calculate an accurate average unless one accounts for all of the outcomes found in the survey.

EXAMPLE 24.5.2 If $X = \{(5,10),(9,3),(27,40)\}$ then we observe that 5 occurred 10 times in the survey, 9 occurred 3 times in the survey, and 27 occurred 40 times in the survey. In calculating the average of X we would include 10 copies of 5 in the sum by writing $10 \cdot 5$. Similar observations apply to $(9,3)$ and $(27,40)$. The total number of outcomes found in the survey would be $10+3+40$. Thus, the mean for this survey is

$$\text{mean of } X = \frac{10 \cdot 5 + 3 \cdot 9 + 40 \cdot 27}{10+3+40}.$$

24.6 Median

Another number associated with the survey is the *median of X*. The calculation of the median of X, written as m_X, requires that we use X or the frequency graph $\Gamma(X)$ of the survey. However, the median of X will exist only for certain samples spaces S. Unless some ordering is associated with the $x \in \mathrm{S}$, the median m_X of X cannot be calculated.

For instance, if we are counting colors or voters then m_X will require some numerical association to the $x \in \mathrm{S}$. Also, unlike the *mean*, if S is a set of words ordered by the usual dictionary ordering, then the frequency graph $\Gamma(X)$ can be drawn with some meaning and a median can be calculated. Hence, before m_X can be calculated, the outcomes $x \in \mathrm{S}$ must be associated with some kind of natural order. For this reason, and because numbers have a well known ordering, we will measure the median m_X of our survey X only when *the associated sample space* S *is a set of numbers.*

The *median of X* is roughly the center of the data X, but not exactly. So let us look at the frequency graph $\Gamma(X)$. Because S is a set of numbers, the *median of X* is precisely that number m_X that splits the bar graph $\Gamma(X)$ into two equal areas. In terms of calculation, *the median of X* is the number m_X that is either an integer N or $N.5$, such that

$$\sum\{n \mid (x,n) \in X \text{ and } x < m_X\} = \sum\{n \mid (x,n) \in X \text{ and } m_X < x\}.$$

See [1]. This equation says precisely that the area to the left of m_X is equal to the area to the right of m_X. This calculation is another reason why we use numbers in S and not other outcomes.

If S is a small set and if the frequencies recorded in X are also small then we can list *as a sequence* the outcomes and their frequencies used to calculate the median of X. The resulting number is erroneously called the "center" of the data.

EXAMPLE 24.6.1 Before we begin, we note that there are far too few data points on which we will do statistics. Meaningful statistical calculations usually require upwards of hundreds of data points. This example does not possess a survey space of such size.

1. If $X = \{(1,4),(2,5)(3,1)(4,1)\}$ then we can represent X as

$$(1,1,1,1,2,2,2,2,2,3,4)$$

and then calculate the median of X as the number at the center of the sequence, $m_X = 2$.

2. If $X = \{(1,1),(2,2)(3,5)\}$ then we can represent X as

$$(1,2,2,3,3,3,3,3)$$

and then calculate the median of X as the number at the center of the sequence, $m_X = 3$.

3. If $X = \{(1,3),(2,2),(3,5)\}$ then we can represent X as the sequence

$$(1,1,1,2,2,3,3,3,3,3)$$

and then calculate the median of X as $m_X = 2.5$.

Sometime after 1975, the examples of medians used in finite mathematics textbooks stopped using sequences containing multiple values of outcomes in their examples of calculations of the median of small sets of data. This misrepresentation of median can be found in books on the history of mathematics [3], books on finite mathematics [5], and even in Webster's Collegiate Dictionary [2]. With this practice, the student is likely to mistake the *center of the sample space* S with the *center of data* X, or with the *median of* X. These are two different numbers, and the examples demonstrating how they are calculated should illustrate these differences.

These examples suggest that the median can be calculated by using the data pairs $(x, n) \in X$ and then ignoring the frequency n of x. There is at least one place in which this method can be used to calculate the median. If you have enough pairs $(x, n) \in X$, and if the frequencies n of $x \in \mathrm{S}$ are large enough then you have an approximation to what is called a *normal distribution.* This is the normal curve that so many faculty try to achieve as they implement their grading policies.

However, unless X is taken from a very special survey, the number of points in X and their frequencies have to be *astronomically large* before the frequency graph can be seen as *approximating* a normal curve. In the usual examples given to students, these huge numbers do not exist, and therefore these examples ignoring frequencies larger than 1 are misleading.

Comment: This is not just a problem in freshmen mathematics texts. This practice of ignoring frequency in the calculation of median can be found in business practices and the implementation of policy.

24.7 Randomness

Let us reexamine the interesting subject of *randomness.*

With the use of today's computers, let us take what is called a *poll* or *survey* of 1000 people. Without further properties, the answers to our survey questions from this 1000 people set, called a *survey sample*, will not tell anyone anything about the qualities possessed by even one more person outside of the 1000 chosen people.

Let us say that the survey sample shows us that 200 out of 1000 are *short*, where short is an abstract and obvious quality possessed by people. If we wish to use our 1000 person survey sample to predict to within a couple of percentage points the number of people in a larger population that are short, then this survey sample must have *randomness*. But first we must know what it means for a survey sample to be *random*.

While the quality of *randomness* is not strange to us, there are only a few who know what it means for a survey sample to be a *random survey sample*. In what might seem to be a bit of circular logic, a survey sample is said to be *random* if it can be used to *exactly predict* qualities of the larger population that the same experiment is designed to measure. That's right, a random survey sample of voters predicts *exactly* how the larger population of voters will vote. A random survey sample of citizens will predict *exactly* how the larger population of citizens will act. Thus, it is difficult for any survey sample to possess the quality called *random*.

Random is such an extraordinary quality of a survey sample that we do not ask for a random survey sample when we collect survey samples. We ask that the survey sample possess *randomness*, which is a significantly different quality than *random*. A survey sample possesses *randomness* if the sample survey *approximates* a random survey sample. The accuracy of the approximation will determine how accurately the survey sample can predict the outcomes of a larger population. This measure of approximation that *randomness* possesses has been made precise through the use of Integral Calculus.

Once we have imbued our survey sample of 1000 people with *randomness* we can predict how many of us *in the larger population* are short to within a small error. That small error is known to you. It is the $\pm 3\%$ that you see in news service polls taken throughout the year.

EXAMPLE 24.7.1 Suppose that we wish to determine how many people in the nation are short through the use of a survey sample of 1000 people. Suppose that our survey sample of 1000 people shows that 20% of the people surveyed are short. This in itself tells us nothing about other populations. But if from the beginning we imbue this survey sample with randomness then we can use this number 20% to predict that within a small error, say 3%, that 20% of the nation will be short. Randomness is that powerful, but a *random* survey sample would exactly predict that 20% of the nation will be short, even though the nation's population is many hundreds of orders of magnitude larger than 1000.

Hence, if we collect a survey sample of 1000 people, and if we insure that our survey sample has *randomness*, then we can use certain subsets of the 1000 to anticipate how larger populations will behave. So if we collect a survey sample of 1000 people, and if our survey sample possesses *randomness*, then from our survey sample's relatively small size we can predict how the entire nation will vote to within a small error. However, while *randomness* is often claimed, it rarely exists in public practices.

Let me give a few examples to illustrate how difficult it is to find *randomness* in surveys.

EXAMPLE 24.7.2 Suppose that the nation is voting on items A and B. Suppose further that an exit poll of a 1000 people voting in Connecticut asks if the person voted for item A or item B. For the purpose of our discussion, let us suppose that 600 people in our survey sample of 1000, or equivalently 60% of the people surveyed, voted for item A.

Those administering the poll gave their pollsters explicit instructions as to how a random choice of people for the survey sample was to be made. These instructions have been modified over many years of experience in polling in Connecticut, so the polling company felt that they would produce a survey sample possessing *randomness*. The pollsters followed their instructions to the letter, producing a survey sample of 1000 that all in the polling company agreed possessed *randomness*.

However it is taken, the above survey sample cannot be a *random sample*. The reason is simple. If said survey sample is *random* then in the sample of the nation's population, we could say with certainty that 60% of the people would vote for item A. Thus, there is no reason to hold that national vote to decide item A or B. All we need do is have a random sample of 1000 people in Connecticut vote, and then the nation's vote on these items would be established with 100% accuracy. This kind of agreement between Connecticut and the rest of the nation has not been seen in the over 200 years of voting in the United States. Since it does not exist in the past, it does not exist with any mathematical reliability. As such, the survey sample is not *random.*

In other words, these 1000 voting Connecticut residents will *never* form a random survey sample in a nation with as many varied points of view as the United States. Specifically, if item A is presidential candidate A and if item B is presidential candidate B then no survey sample of 1000 people in Connecticut will be random since Connecticut has never determined the nation's vote on its future president. No survey sample collected in this way can legitimately be called *random.*

But wait, there's more.

EXAMPLE 24.7.3 Assume to the contrary that our survey sample of 1000 voting Connecticut residents possesses *randomness.* Then there is a small error, say 3%, such that 60% $\pm$ 3% of the people in the nation voted for item A. The assumed *randomness* of our survey sample would imply that at least 57% of the nation would vote for item A, and this kind of agreement has not ever surfaced in the voting process in the United States. This historical proof of this mathematical impossibility shows that no survey sample of 1000 voting Connecticut residents will possess the *randomness* necessary to predict the voting habits of the rest of the nation. Subsequently, exit poll surveys taken at polling stations on the East Coast are not likely to influence voting habits in California *unless we allow them to do so.*

But wait, even more is true.

EXAMPLE 24.7.4 Assume that the nation is filling in ballots whose lone question will be answered by one of item A or item B. A survey sample of 10 million people is taken from the east coast states which supposedly possesses randomness. It is found that 100% of those 10 million sampled chose item A. From the randomness supposedly possessed by this survey sample, we would conclude that 100% of the nation will vote with item A. But a curious thing happens at this point in our story. When the results of the survey are published, a similar sample of 15 million people from the west is published, and it reveals that 100% of the 15 million voters sampled have circled item B on there ballot. The reason for this disparity is quite simple.

It seems that the ballot is a census form and not a vote on a political issue. The lone question on the ballot is *If you live east of the Mississippi then circle item A. If you live west of the Mississippi then circle item B.* Our hope that our 10 million person survey possesses randomness is utterly crushed when we read that 100% of the west coast voters disagree with our 10 million person survey.

Therefore, no small localized survey sample can possess randomness. Moreover, the previous example shows that a knowledge of questions on the survey is a must if we are to find randomness in our chosen survey sample.

The question of *randomness* in a survey sample may not strike you as difficult given all of the surveys and polls taken and reported for your consumption on the evening news. You are bombarded with one 1000 person survey sample after another, so you might expect that *randomness* in a 1000 person survey sample is quite common. But that would be wrong, as the above examples show.

Indeed, when a Polling Company states that their survey was taken from 1000 people chosen at random, and when that use of *random* means that they polled people coming out of a local mall in Iowa, you can be sure without error, that survey sample did not possess *randomness.*

It takes time and money to collect a survey sample possessing *randomness.* The choices made require that the people chosen for the survey sample satisfy a long list of qualities before they can belong to a survey sample possessing randomness. The ages, the races, income brackets, property ownership, political affiliations, height, and many more human factors have to be considered when choosing an approximation to a survey sample that possesses *randomness* that can predict statistical behavior from a larger population.

There are those in government circles who have the resources to produce a survey sample possessing *randomness* on a regular basis. From such a survey sample a government can tell how the entire population of the nation thinks or feels about controversial issues, to within a small percentage error.

If you try to improve the *randomness* in your survey sample by simply making it larger, you would also be in the wrong. *More* people will not give a survey sample *randomness*. In fact, size doesn't matter here. After all, you cannot tell how the French will vote on item A by taking a survey of American voters. To produce *randomness* you do not need more people in your survey, you need more *randomness*. These are two different things. To produce *randomness* in a survey, a better method of polling is required. The most frequently practiced error is to think that *more* is more accurate when thinking about survey samples.

Chapter 25

Linear Programming

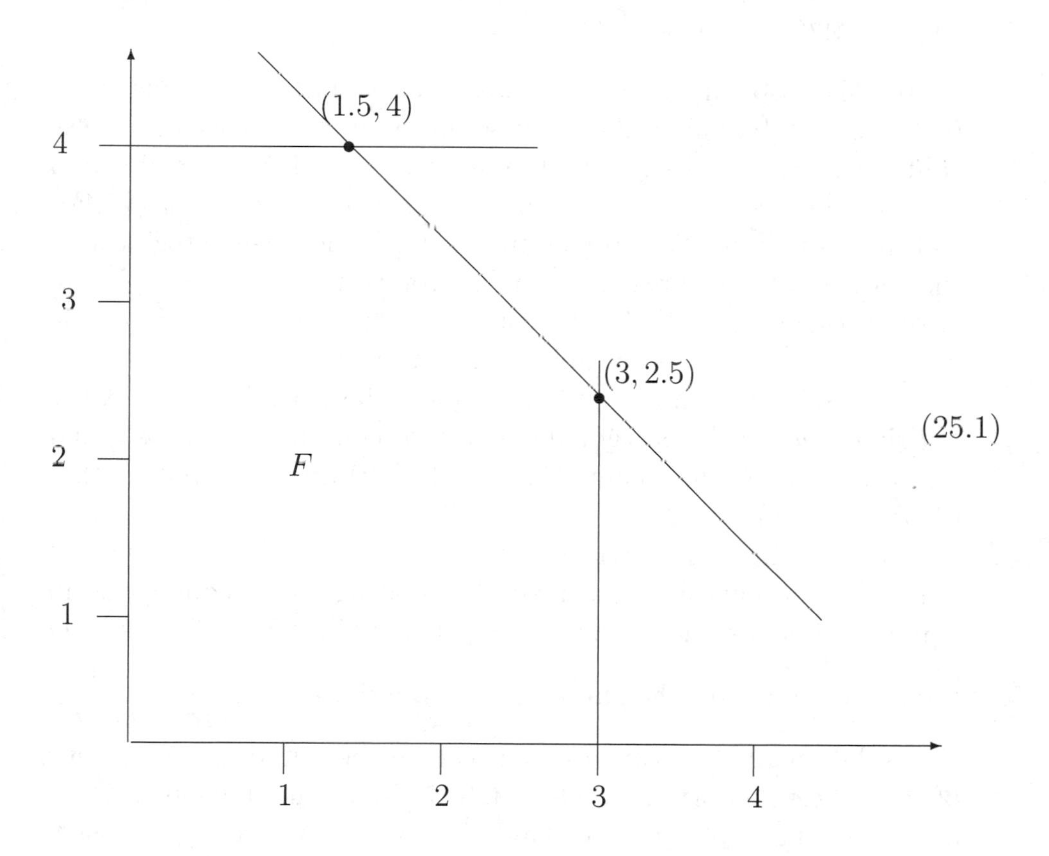

(25.1)

We examine the geometric linear programming problems generally taught in the freshmen finite mathematics course, showing that the methods used are not always correctly applied.

25.1 Continuous Variables

There is a word problem generally taught in finite mathematics courses that goes something like the examples below.

EXAMPLE 25.1.1 Given a number $N > 0$, and variable quantities $0 \leq x,\ y \leq N$ that cost A and B cents, respectively, find a pair (u, v) such that the cost $Au + Bv$ is as small as possible with respect to a finite set $\mathbf{D}$ of inequalities of the form $Ux + Vy \leq 1$ for some numbers U and V. The set of points (x, y) in the plane that satisfy the inequalities in $\mathbf{D}$ form a polygonal region F called the *feasible region* of the problem.

In this problem, the values chosen for x and y are nonnegative real numbers bounded above by some number N. Because these values x and y can be any real number less than N, we call x and y *continuous variables* or *continuous variables bounded by* N. For continuous variables x and y bounded by N, the feasible region F is the convex polygonal region, including its interior. A typical picture of the *feasible region* F of the linear programming problem Example 25.1.1 with continuous variables is given in (25.1).

Because the variables are continuous, the feasible region is the *entire interior* of the polygonal region that contains the letter F and its border. Thus, the points $(1, 1)$ and $(3, 2)$ are in F, but neither $(3, 3)$ nor $(4, 1)$ is in F.

One identifies the vertices of the boundary of F by solving the pairs of linear equations that result by treating each inequality as an equation. For instance, if $Ux + Vy \leq 1$ and $U'x + V'y \leq 1$ are in $\mathbf{D}$ then we would solve the pair of linear equations $\begin{cases} Ux + Vy = 1 \\ U'x + V'y = 1 \end{cases}$.

The solution to this pair of equations can be found using any one of several elementary methods taught in high school algebra. Once all of these pairs of equations have been solved we are in possession of the corners of, or *the vertices*, of the feasible region F.

If a point (x, y) is not in F then (x, y) does not satisfy some inequality in $\mathbf{D}$, and so (x, y) will not be a solution to our problem Example 25.1.1. The feasible region F then contains all of the potential solutions to the problem. *Thus, if a solution exists, we* *will find it in F.* In other words, to solve the linear programming problem, the located point (u, v) that minimizes the cost must be in the feasible region, F, of the problem. If the point chosen is not contained in F then it is not a solution to the problem.

We will call variables $x > 0$ and $y > 0$ whose values are taken from the set $\mathbb{N}$ of natural numbers *discrete variables.* When any freshmen finite mathematics text of the last 50 years is examined, we find that continuous variables are not used in linear programming problems presented as word problems or as applications. These problems contain variables that take on *discrete* values from the set of integers, $\mathbb{N}$. Problems with continuous variables and problems with discrete variables are quite different and require different methods of solution. The next example will demonstrate.

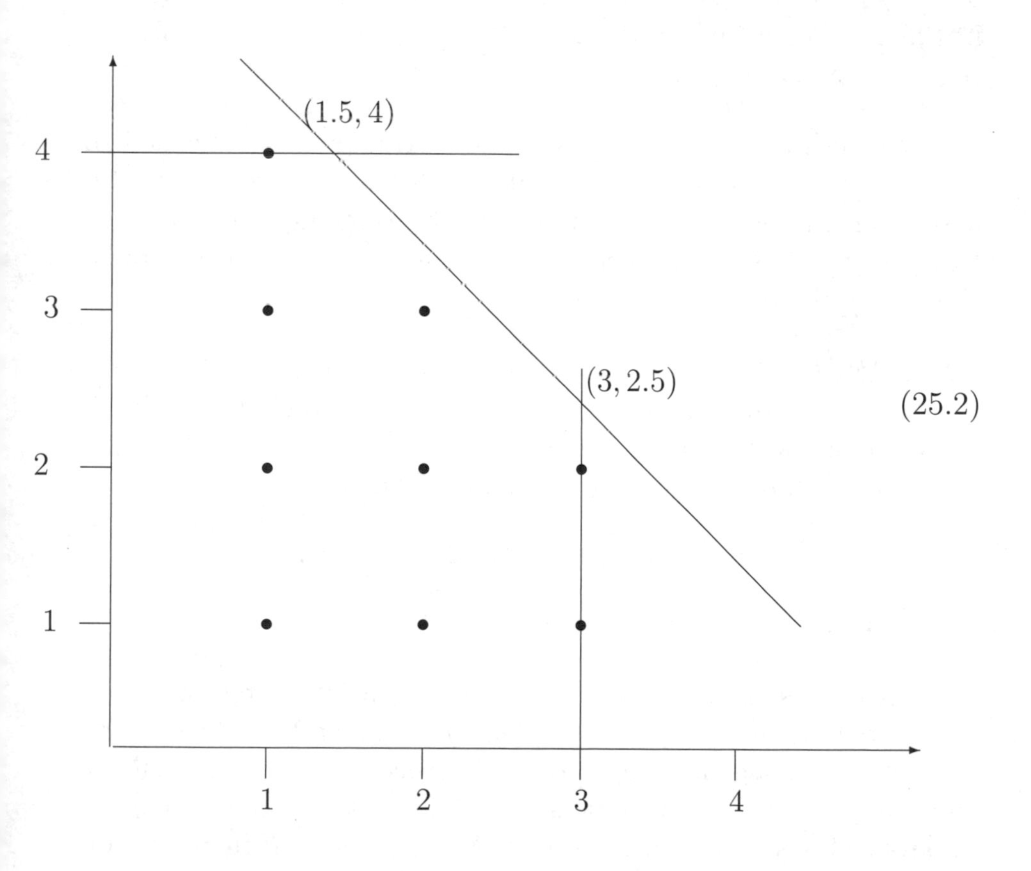

(25.2)

EXAMPLE 25.1.2 Let $x > 0$ be the number of widgets sold at 20 cents a widget, and let $y > 0$ be the number simple machines sold at 30 cents a machine. Because physical space is limited, there can be no more than 4 widgets stored, and no more than 3 machines stored. The number of widgets and machines is related by $2(x + y) \leq 5$. Find the number of widgets and machines that will minimize the cost $C = 20x + 30y$.

From this word problem we can write down the set of inequalities

$$\mathbf{D} = \begin{cases} 2x + 2y \leq 5 \\ 0 < x \leq 4 \\ 0 < y \leq 3 \end{cases} .$$

The inequalities in $\mathbf{D}$ translate into three linear equations $2(x+y) = 5, x = 4, y = 3$, which define the borders of the feasible region F in (25.2). We will minimize $C = 20x + 30y$ relative to the feasible region graphed in (25.2).

The method that we will use to solve this problem is called the *ruler method* because the ruler plays a significant role in finding a solution in the feasible region F. The ruler method as taught today first draws the *feasible region* F in the first quadrant of the Euclidean plane, as we have done in (25.2).

EXAMPLE 25.1.3 Let us continue to solve the problem in Example 25.1.2 by finding the vertices of the polygonal region F. Solve for the vertices of F by solving the pairs of equations $x, y > 0, 2(x+y) = 5, x = 4, y = 3$ defined by the inequalities $\mathbf{D}$. Then,

$$(0, 0), (0, 4), (1.5, 4), (3, 2.5), (3, 0)$$

are the vertices of F.

The next step in our method of solution is the reason why it carries the name of *the ruler method.* In a general, this melding of geometric and algebraic optimization processes will use a ruler to continuously and smoothly translate or move the line $Ax + By = C$ in parallel lines until it intersects the feasible region F in exactly one of its vertices (u, v). The legend goes that this intersected vertex (u, v) is the point at which $Au + Bv = C$ is the minimum value of the value C over all possible points $(x, y) \in F$.

Let us complete Example 25.1.2 by using a ruler to find the vertex at which the minimum cost is achieved.

EXAMPLE 25.1.4 You can solve the problem in Example 25.1.1 by using a ruler to draw a line parallel to $20x + 30y = C$ that intersects F at the vertex $(1.5, 4)$. This is your ruler method solution to the linear programming problem in Example 25.1.2.

A more systematic solution to this problem is to check $Ax + By$ over all vertices (x, y) of F, and choose the pair (u, v) that results in the smallest cost $C = Au + Bv$ over all *vertices*. This will also result in the minimum cost over all pairs $(x, y) \in F$. We will also call this method the *ruler method.*

Then you can solve the problem in Example 25.1.1 by substituting the vertices (u, v) of F into $20x + 30y = C$, and save the one that yields the least value of C. Observe. Check the values of C for the vertices $(1.5, 4)$ and the other vertices. Then $C(1.5, 4)$ will be the least value of C when compared with the values $C(3, 2.5), C(0, 0), C(0, 4)$, and $C(3, 0)$. Thus, $(1.5, 4)$ yields the minimum value of C over the vertices of the polygonal region F.

REMARK 25.1.5 Despite a generally held opinion, the pair $(1.5, 4)$ *does not solve the problem given in Example 25.1.2.* (Note the change.) Remember, by the hypotheses any solution to the linear programming problem given in Example 25.1.2 must feature nonnegative **integers** $x \leq 4$ and $y \leq 3$. If x is not an **integer**, then (x, y) is not a solution to Example 25.1.2. Since $x = 1.5$ in our chosen point above, $(1.5, 4)$ is not a solution to Example 25.1.2. It does not solve the given problem because it fails to satisfy the problem's hypotheses. We have instead solved some other problem.

Indeed, since the pairs (x, y) in a solution must be a pair of integers, only the lattice of points drawn in the graph (25.2) can be considered when solving the linear programming problem in Example 25.1.2. It follows that *the feasible region for our linear programming problem is a finite set of points* and not the polygonal region F graphed in (25.2).

25.2 Discrete Variables

We will continue to examine the linear programming problem Example 25.1.2.

The conclusions we drew in Remark 25.1.5 show that while the *ruler method* might work for continuous variables x and y of nonnegative real numbers bounded by a number N, it does not in general yield a solution to linear programming problems whose variables are discrete values. The ruler method then possesses a problem as it does not perform as advertised.

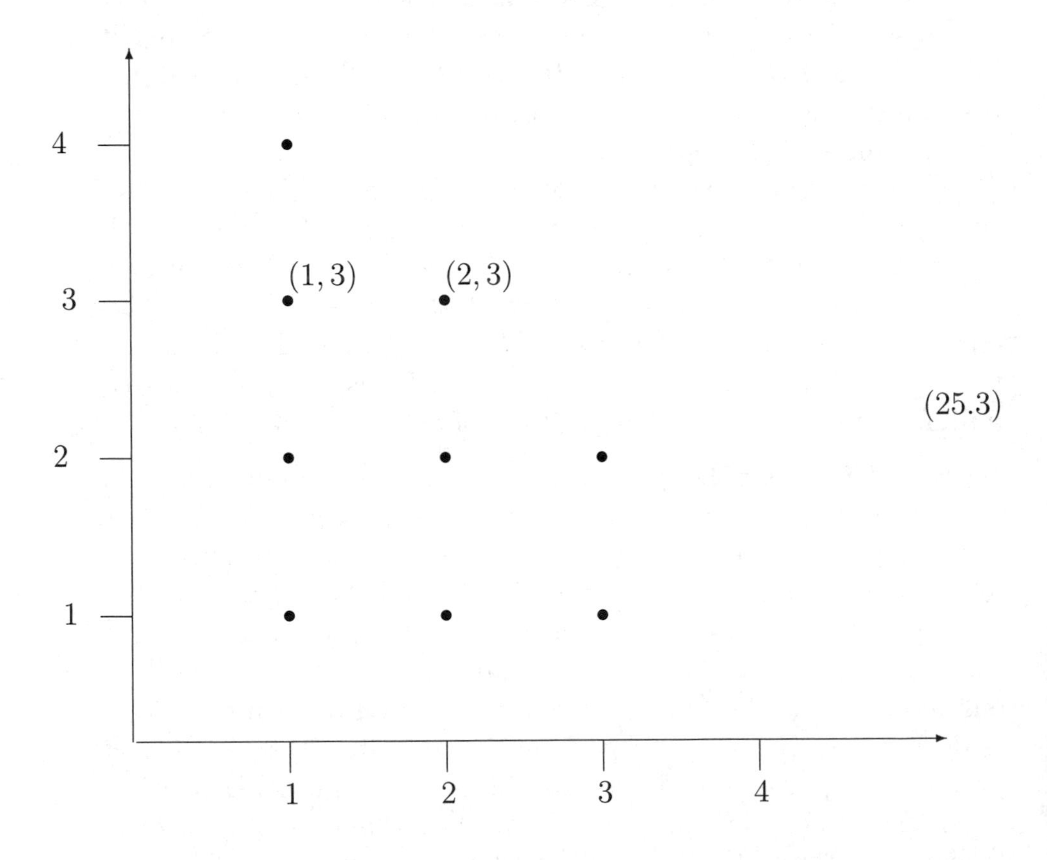

(25.3)

Examples of discrete variables include those variables x and y that represent the numbers of cars and trucks observed, or the number of jars and cardboard containers shipped, or perhaps x and y represent the number of thing 1 and thing 2 that are sold over a holiday weekend. In each of these examples, the only values taken on by the variables x and y are nonnegative integer values.

Observe that the variables in Example 25.1.2 are discrete variables, because they take their values from the set $\{0, 1, 2, 3, 4\}$. The following graph shows the points in the feasible region of the linear programming problem in Example 25.1.2. Compare these points to the polygonal region F given in (25.1).

The vertices of the polygonal region F do not appear in this diagram because they are not integer pairs that satisfy all of the inequalities in Example 25.1.2. The solution to our linear programming problem must be a pair of integers, so only integer pairs will appear in our feasible region. We observe that this feasible region is a finite set of points and not a polygonal region.

Let E denote *the finite lattice of points graphed in (25.3).* Then E is a finite set of integer pairs (x, y) contained in F, and as such they solve the linear programming problem given in Example 25.1.2. It is then natural to call E the *feasible region* of linear programming problem given in Example 25.1.2, even though it is not a polygonal region in the plane.

We will assume for the remainder of this section that x and y take on integer values between 0 and N. Specifically, there can only be a finite set of points (x, y) to consider in our solutions to the equation $Ax + By = C$ and the inequalities in $\mathbf{D}$. We will also say that (x, y) is an *integer point bounded by* N.

EXAMPLE 25.2.1 In order to be a solution to $Ax + By = C$ in Example 25.1.2, the point (x, y) must be an integer pair bounded by N. Otherwise, it is not a viable solution. So, when we use that ruler to continuously translate the line $Ax + By = C$ through parallel lines until it intersects exactly one vertex of E, we must disregard the lines parallel to $Ax + By = C$ that are solved by *non-integer pairs* (x, y).

Therefore, when solving the linear programming problem in Example 25.2.1, which is defined using discrete variables, the ruler method must ignore almost all of the lines described by the ruler, and instead use only those lines for which C is one of the accepted non-negative integers.

Evidently, the ruler has no way of keeping track of which points on the continuous line are actually integer pairs bounded by N. So when the continuous line $Ax + By = C$ intersects with a vertex in the polygonal region F, there is no geometric way to determine which points on that continuous line are some of the finitely many integer pairs that are a part of this problem. The ruler method does not tell us which pairs on the line $Ax+By = C$ are integer pairs in E, but these are the only pairs that we can consider in solving the linear programming problem in Example 25.1.2.

EXAMPLE 25.2.2 This is our principle mathematical difficulty with the ruler method when it is used to solve the linear programming problems like the one in Example 25.1.2. It is unlikely that the one point intersection (u, v) between the continuous line $Ax+By = C$ and the polygonal region F is *an integer pair*. Hence, it is unlikely that (u, v) is a solution to the linear programming problem in Example 25.1.2 as our solutions must be integer pairs.

We employed the ruler method with the confidence that it will produce a solution to the problem. It has not. A method of solution should reliably find a solution to the problem to which it is being applied.

EXAMPLE 25.2.3 Another significant objection to the use of the ruler method to solve linear programming problems like Example 25.1.2 is that in every finite mathematics text since 1960, each example and each exercise in the form of a **word problem** features feasible regions F, each of whose vertices are integer pairs. If one were to use this spotty research then the conclusion might be that integer variables always imply integer pairs as vertices in the feasible region. This conclusion is inaccurate of course, as figure (25.1) verifies.

Of the 15 finite mathematics texts on my shelf, not one of them offered a linear programming **word problem** with integer variables, and that featured a non-integer pair as a vertex in F. Each of these plentiful problems are so designed that the vertices of the feasible region are integer pairs, and as such they are potential solutions to the stated problem. Perhaps these integers owe their existence to clever professors who wanted the arithmetic in the assigned homework problems to be as easy as possible.

EXAMPLE 25.2.4 The examples and exercises in these texts on linear programming problems represent a stacked deck of sorts, as the method used to solve them only works on the problems in the book. Because the authors have published word problems for which the vertices of the polygonal region F are integer pairs, we are confronted with a difficult to believe coincidence. This apparent impossible accident of integer vertices does not show the reader how to implement the ruler method on problems outside of the class.

As it is near certain that some misinformed individual will misuse the ruler method to incoprrectly solve a problem while on the job, I suggest that *the ruler method serves no useful purpose* on the job or in the classroom.

EXAMPLE 25.2.5 Let P be a linear programming problem whose variables are integers x and y bounded by N, and suppose that the polygonal region F in (25.1) and (25.2) contains the integer pairs that are the solutions to P, and suppose further that the ruler method has produced the point $(3, 2.5)$ to solve P. The pair $(x, y) = (3, 2.5)$ might solve P *over all real numbers*, but it does not solve our problem since any solution (x, y) to our problem must be an integer pair in F.

The current thinking in some freshmen finite mathematics texts is that we can make a solution of $(3, 2.5)$ if we change $(3, 2.5)$ to the point $(3, 3)$. There is a fatal flaw in this freshmen reasoning, though. Since $(3, 2.5)$ is a vertex on the polygonal region F, $(3, 3)$ *is not in* F. Because $(3, 3) \notin F$, $(3, 3)$ fails to satisfy at least one of the inequalities in $\mathbf{D}$. Our solution to P must be in F, and it must satisfy all of the inequalities in $\mathbf{D}$.

More important is the fact that once we have changed from $(3, 2.5)$ to $(3, 3)$ *we have exited the proven algorithm that is used to solve these linear programming problems.* The idea of changing the pair $(3, 2.5)$ into a pair $(3, 3)$ that will still yield a smallest value for $Ax + By$ is wishful thinking and not mathematics.

Inasmuch as there is no apparent logic to justify the assumption that $(3, 3)$ will yield a smallest value for $Ax + By$, we must now compare $A3 + B3 \leq Ax + By$ where (x, y) range over all integer pairs (x, y) in F. Anything less than this effort will not reliably produce an integer pair in F at which the minimum value for C is achieved.

A deeper criticism of the ruler method is that no one seems to know a proof that shows that the ruler method works. Until this demonstration of correctness has been done, the ruler method is incorrectly applied to linear programming problems.

EXAMPLE 25.2.6 The most important comment that can be made about solutions to linear programming problems with integer variables, though, is that is that *the vertices* $(1.5, 4)$ *and* $(3, 2.5)$ *of* F *are not vertices of* E. Thus, we need to know which points in E we should test in solving the problem.

Furthermore, there are good arguments that the points $(1, 3)$ and $(2, 3)$ are vertices of E. Vertices worked before as solutions to linear programming problems, so it would be advantageous to know what a vertex is in E. Once the term *vertex* is defined, we need an efficient method for finding them.

In any event, we need to know which of the points in E we should test in our search for a solution to the linear programming problem. We ought to be prepared, though, for the possibility that the minimum solution does not occur at a newly defined vertex of E.

25.3 Incorrectly Applied Rules

In this section we will show in one example what happens when someone misapplies or ignores existing rules in arithmetic. The result is a tongue in cheek slap at those who fail to follow mathematical rules when doing mathematics.

A young sailor walks into a bakery and says that he has to purchase donuts for each officer on board his ship. He has counted out the problem and decided that if he is to give 7 donuts to each of the 13 officers then he must purchase 28 donuts total. Here is his arithmetic reasoning and the method he used to check his answer.

He performs the first step like a fourth grade graduate by multiplying 3 by 7 to come up with 21. $\begin{array}{r} 13 \\ \times \quad 7 \\ \hline 21 \end{array}$. He then completes the multiplication by multiplying 1 by 7 and adding the two numbers together as follows. $\begin{array}{r} 13 \\ \times \quad 7 \\ \hline 21 \\ 7 \\ \hline 28 \end{array}$. This is his anticipated 28. He checks his answer with addition. First he sums up the right-hand column of the sum $\begin{array}{r} 13 \\ 13 \\ 13 \\ 13 \\ 13 \\ 13 \\ 13 \\ \hline 21 \end{array}$. Then the 1's are added as 1 and 1 and 1 and 1 and 1 and 1 and 1, which he believes is equal to 7. This he adds to 21 as $\begin{array}{r} 13 \\ 13 \\ 13 \\ 13 \\ 13 \\ 13 \\ +\ 13 \\ \hline 21 \\ 7 \\ \hline 28 \end{array}$. He is satisfied that he must purchase 28 donuts.

He needs one more check of his count, so he will divide 7 into 28 in the hopes of getting 13. He writes $7\overline{)28}$ and argues thusly. “Seven will not go into that little 2 no matter how hard you try. So let me take that 2 from the paper and give it to you to hold until I call for it.” The 2 is taken, leaving an unadorned 8. $7\overline{)8}$ “Seven,” he calculates, “goes into that 8 just once.” He writes down his result $\begin{array}{r} 1 \\ 7\,\overline{)\,8} \\ 7 \\ \hline 1 \end{array}$. "Now I need that 2 that I asked you to hold.

Give it to me, please." The 2 is taken and he writes $\begin{array}{r} 1 \\ 7\,\overline{)\,8} \\ 7 \\ \hline 21 \end{array}$. His eyes pop open as he sees that 7 goes into 21 exactly 3 times with no remainder. He writes $\begin{array}{r} 13 \\ 7\,\overline{)\,8} \\ 7 \\ \hline 21 \\ 21 \\ \hline 0 \end{array}$. Exalted, he purchases and delivers his 13 donuts.

The Moral: The ruler method is not the first time that someone has misused the rules of mathematics.

Chapter 26

Subjective Truth

Theorems, lemmas, propositions, and corollaries are mathematical statements that are the conclusion of a series of implications collectively called an *argument* or a *proof*. Prior to 1970, mathematics was known for the absolute Truths that were the conclusions of its proofs. Recently, a group of pseudo-intellectuals have tried to present to the world their misinformed belief that mathematical Truths are not universal Truths, or *absolute* Truths. They claim the Truths of mathematics are *subjective*, and that these Truths are open to discussion.

26.1 The Absolute Truth of an Axiom

A proof follows a specific pattern that was first discovered and subsequently laid down by Aristotle some 2600 years ago. Each proof in mathematical discourse could, theoretically but not in practice, begin with a subset of a fixed set of primitive accepted Truths called *Axioms*, and then proceed using Aristotle's classic arguments to a conclusion that would also be True. The mathematicians of the day were experts at determining which proof represented absolute Truth, and which did not. If we read an absolute proof then its conclusion is an absolute Truth.

An *Axiom* is a statement that is accepted as True within a given logical system. A *model* for a given set of Axioms is a mathematical set in which the Axioms are True. *Within the model the Truth of these Axioms is to be accepted without qualification.* You were free to change an Axiom and you could introduce a new Axiom, but then the model associated with your Axioms *must also change.* Your new model would be different from the model you were working with just before you changed Axioms. Thus, it is important that mathematics has one agreed upon set of Axioms, and therefore exactly one fixed model. In the early twentieth century, mathematicians agreed that the ZFC Axioms of Set Theory were the best available set of Axioms for modern mathematics. These are the primitive Truths that mathematics can be reduced to if such a reduction was found necessary. Thus, Set Theory is the basis for the most important of mathematics discussed today.

Other areas of mathematics begin with the ZFC Axioms of Set Theory and additional Axioms peculiar to that area. For example, from Euclid's Five Postulates (=Axioms), most of plane geometry is deduced, from the Peano Postulates, most of arithmetic is deduced, and from the Axioms of algebra, most of algebra is deduced. From these Axioms we prove the absolute Truth of most of the more advanced ideas in mathematics. While there are statements in mathematics that cannot be deduced from the ZFC Axioms, these statements are so far removed from the undergraduate courses in mathematics that we can ignore them here.

Sometime after 1970, a group of pseudo-intellectuals began to publicly introduce *subjective Truth* to rational discussions. The notion of a subjective Truth was not defined in their discussions, nor could they be pressured to inform the listeners when they were exercising subjective Truth. Thus, a frontal attack on their methods was difficult to construct.

By reading the arguments written by this group one gets the impression that Truth to them is a flexible thing. You can assume a statement Q or its logical negation $\neg$Q, and you would be right in either case. According to them, you would be right because Truth is just an opinion, and not something written in ancient stone. It seemed to me that Truth to them, just like the Truth of an opinion, is a function of how the majority votes.

In my surfing on the web during the period 2002 through 2007, I came upon a group of these pseudo-intellectuals at mathematical and scientific threads. When I introduced them to a short argument that *All is known* is a Falsehood, I was met with a series of postings that in the end became a lengthy personal attack. Those enraged responses to my logical argument made me feel a bit uneasy, as I was used to rational discussion when two intellectuals disagree. Evidently, these people were not members of the collegiate community. Be warned that in those days, this was the common state when one introduced correct logical argument to these threads.

When I posted a proof that there were no Universal Turing machines, the responses I read were so weak on collegiate disagreement that I found it hard to read the responses. Given their passionate postings, it seemed to me that most of those replying felt that the existence of a Universal Turing machine was a religious thing. Obviously, my opinion on the subject was not recognized.

When I posted a proof that the collection **C** of all sets is not a set, I was met again by groups that seemed to hold **C** in high religious esteem. Now how do you argue with that? One man's response to my construction of the natural numbers from the empty set stated that God created the natural numbers, not me. I replied that the statement *God created the natural numbers* was actually first made by a mathematician Kronecker in the 1880's, when he read that a peer Cantor had found numbers called cardinals of infinite value, [4]. The statement was meant to be a jibe at the questionable parentage of infinite cardinals. My post received no replies.

There was one person who reported that 350 faculty from some fringe think tank had signed a paper declaring that Darwin was entirely incorrect. My reply quoted Einstein. *Just one would suffice if he had a solid argument.* And finally, when I suggested that mathematics deals in absolute Truth, I was rebuffed by a group who claimed that Truth was subjective. This sounded to me like they had mulch for brains. At this point I cast aside my blue mood, as I was possessed to learn more.

Let us examine some examples of their exposed thought processes. Suppose that I say that the statement *The sky is blue* is an absolute Truth. The pseudo-intellectual would argue that *The sky is blue* is False if he is on Mars. This, he would claim, is evidence that Truth is subjective. The role of *model* is completely ignored here. One cannot say he has a Truth unless he first specifies where he is or what *model* he is using for the discussion. *The sky is blue* is an ambiguous statement if it is not made within a model, and without such a model one cannot attach a True or False value to it.

When we say *The sky is blue* what most of us mean is the statement *The sky is blue above me on Earth.* The position is implied because most of us live on Earth, and not on Mars. In this case, no ambiguity occurs because *The sky is blue above me on Earth* is meteorologically True. Once one sits on Mars *The sky is blue above me on Earth* is still True. Then this Truth seems to be absolute. No matter where we are, *The sky is blue above me on Earth* is True.

Mathematics provides the next example. As was said above, an Axiom *for a specific model* is accepted as an absolute Truth. The pseudo-intellectuals would respond by saying that *the Truth of anything is always subject to discussion*, completely missing the definition of an Axiom. An Axiom, they would answer, is just an *opinion*, and any opinion offered as an alternative axiom is *valid.* They conclude that an Axiom is not an *absolute* Truth.

Their conclusion erroneously confuses *validity* with *Truth.* Also, they have ignored the fact that the Truth of a mathematical Axiom *within a fixed model* is not a subject for debate. In arguing so, these pseudo-intellectuals would be working from one of the most subtle of Falsehoods ever written. The argument given in the previous paragraph is based on the belief that *All opinions are valid.* Those who believe this will not tell you what an opinion is, and they will not tell you what the word valid means, but they believe that they have won the argument once they talk about valid opinions. This is wrong of course. Let me show you why.

Assume for the sake of contradiction that *All opinions are valid.* Let Q be the statement *This opinion is not valid.* By assumption, all opinions are valid, so the opinion Q is valid. On the other hand, the content of Q states that the opinion Q is not valid. We have deduced *Q is valid* and its logical negation *Q is not valid*, a contradiction. Thus, *All opinions are valid* is False. When I posted this argument, I was met by two people whose posted responses were so angry and insulting that their postings could be seen as a declaration that my opinion on valid opinions was not valid. I thanked them for supporting my point of view. No one replied.

Suppose a mathematician wrote that Euclid's parallel postulate was an absolute Truth in plane geometry. This is not anyone's opinion. It is a statement that defines the model in which it is applied. We might, as professionals, change the statement of Euclid's parallel postulate to an equivalent Playfair's Axiom, but in doing so we have not dismissed the Truth of Euclid's parallel postulate.

If we change the parallel postulate to something new then we are no longer working within plane geometry. The introduction of different Axioms results in a change from the plane geometry model to something else. The parallel postulate is still True in its model of plane geometry, so no Truth in plane geometry has been unseated. Thus, the parallel postulate remains an absolute Truth within plane geometry. The new axioms simply apply to a different model. This is a common error among the pseudo-intellectuals who practice subjective Truth. They ignore any and all definitions of Axiom, model, and Truth because it is convenient to do so.

The pseudo-intellectual would argue that he could change from professionally written Axioms to statements of another kind and still have a geometry. The chances of such a random change in the Axioms resulting in a geometry are slight. And if he were to try to apply his statements to the model for plane geometry, namely the plane, he would be entirely wrong. By changing the given Axioms, he cannot use the model for plane geometry as a model for his new geometry. A model must be found that will match his axioms, thus giving him a setting in which he can apply the "geometry" to which his axioms will lead.

If the pseudo-intellectual wants to change the Axioms being used, he is welcome to try but in so doing he would change the underlying model and thus change the setting in which the mathematics could be applied. Change the Axioms and you change the universe in which these Axioms exist. You did not replace a Truth with another Truth, you changed models, thus rendering all of the original Axioms inapplicable. The original Axioms were still accepted Truths within their model M, and your new set of axioms could be accepted as Truths within their model N, but by changing Axioms the basic models have changed as well. No Truth was overturned by changing models, as Truth depends on the model in which you are working.

Bibliography

[1] M. G. Blumer, Principles of Statistics, Dover Publications, Inc., New York (1979).

[2] D. B. Guralnik, *Editor in Chief*, Webster's New World Dictionary of the American Language, second college edition, Simon and Schuster (1982).

[3] J. H. Eves, An Introduction to the History of Mathematics, sixth edition, Saunders College Publishing, Chicago (1990).

[4] T.G. Faticoni, The Mathematics of Infinity: A Guide to Great Ideas second edition, John Wiley & Sons, Inc., New York (2012).

[5] M. L. Lial, R. N. Greenwell, N. P. Ritchey, Finite Mathematics, eighth edition, Pearson, Addison-Wesley, Boston (2005).

Index